THE SEA

The Earth series traces the historical significance and cultural history of natural phenomena. Written by experts who are passionate about their subject, titles in the series bring together science, art, literature, mythology, religion and popular culture, exploring and explaining the planet we inhabit in new and exciting ways.

Series editor: Daniel Allen

In the same series

Air Peter Adey
Cave Ralph Crane and Lisa Fletcher
Clouds Richard Hamblyn
Coal Ralph Crane
Comets P. Andrew Karam
Desert Roslynn D. Haynes
Earthquake Andrew Robinson
Fire Stephen J. Pyne
Flood John Withington
Glacier Peter G. Knight
Gold Rebecca Zorach
    and Michael W. Phillips Jr
Ice Klaus Dodds
Islands Stephen A. Royle
Lightning Derek M. Elsom

Meteorite Maria Golia
Moon Edgar Williams
Mountain Veronica della Dora
North Pole Michael Bravo
Rainbows Daniel MacCannell
The Sea Richard Hamblyn
Silver Lindsay Shen
South Pole Elizabeth Leane
Storm John Withington
Swamp Anthony Wilson
Tsunami Richard Hamblyn
Volcano James Hamilton
Water Veronica Strang
Waterfall Brian J. Hudson

The Sea

Richard Hamblyn

REAKTION BOOKS

For Ben and Jessie

Published by Reaktion Books Ltd
Unit 32, Waterside
44–48 Wharf Road
London N1 7UX, UK
www.reaktionbooks.co.uk

First published 2021

Printed and bound in India by Replika Press Pvt. Ltd

A catalogue record for this book is available from the British Library

ISBN 978 1 78914 487 1

CONTENTS

Introduction: 'The Sea Is Like Music'

In civilizations without boats, dreams dry up.
Michel Foucault, *Heterotopias* (1967)

In late August 1909 Sigmund Freud and Carl Jung crossed the Atlantic together in a steamship as the invited guests of a recently founded American university. The six-day crossing from the port of Bremen had a particular impact on the 34-year-old Jung, who declared in his journal that 'the sea is like music: it has all the dreams of the soul within itself and sounds them over.'[1] The pair was entranced – 'the beauty and grandeur of the sea consists in our being forced down into the fruitful bottomlands of our own psyches' – and at the end of the lecture tour, they were impatient for the return journey: 'I am looking forward enormously to getting back to the sea again,' wrote Jung, 'where the overstimulated psyche can recover in the presence of that infinite peace and spaciousness.' The voyage home, though it proved far from peaceful, did not disappoint:

> Yesterday there was a storm that lasted all day until nearly midnight. Most of the day I stood up front, under the bridge, on a protected and elevated spot, and admired the magnificent spectacle as the mountainous waves rolled up and poured a whirling cloud of foam over the ship. The ship began to roll fearfully, and several times we were soaked by a salty shower. It turned cold, and we went in for a cup of tea.[2]

Claude Monet, *Cabin of the Customs Watch*, 1882, oil on canvas, an image about the act of looking out to sea, about keeping watch, an activity for which a large proportion of coastal infrastructure was built over the centuries.

For the rest of his life, the sea remained the Swiss-born Jung's foremost symbol for the deep unconscious, a site of fascination and fear, of grandeur and dread, a threshold that can never be lightly crossed. 'Of all the jumbles of matter in the world the sea is the most indivisible and the most profound,' observed Victor Hugo, in a statement that expresses one of the ruling ideas of this book, although a volume as short as this can only hope to touch the surface of such a vast and unfathomable subject.[3] The sea is in all ways immense – physically it covers more than 70 per cent of Earth's surface, while conceptually it overwhelms the human imagination – and it would be impossible for a book on this scale to offer more than the briefest historical outline of the sea and its multiple meanings.

The chapters that follow seek to trace a cultural and geographical journey from estuary to abyss, beginning with the topographies of the shoreline and ending with the likely futures of our maritime environments. Along the way they will consider the sea as a site of work and endurance, of song and story, of language, leisure and longing, of peace and war. The sea has always acted as a conduit for culture, from the early Pacific voyagers in their outrigger canoes to the Renaissance circum-navigators of the world, and on into our own time, where the seas remain global trade routes, with 90 per cent of the world's commercial goods transported on container ships that cross from port to port via unregulated waters, far beyond national borders, amid the forgotten spaces of the ocean.[4]

'Seas', 'waters', 'oceans': some definitions are called for. 'Seas' are distinct from 'oceans', being the complex regions where land

Winslow Homer, *The Beach, Late Afternoon*, c. 1870, oil on wood. A contemplative view of a Massachusetts beach, painted shortly after the artist's return from France.

and ocean interact. Seas thus behave (and are experienced and understood) in different ways from oceans; to borrow Captain W. H. Smyth's definition, from his *Sailor's Word-Book* (1867), 'strictly speaking, *sea* is the next large division of water after *ocean*, but in its special sense signifies only any large portion of the great mass of waters almost surrounded by land, as the Black, the White, the Baltic, the China, and the Mediterranean seas, and in a general sense in contradistinction to land.'[5] The world's five oceans, by contrast (Pacific, Atlantic, Indian, Arctic and Southern) really join up as one vast saltwater body, the singular surrounding Ocean, in the margins of which the seas are located, both geographically and figuratively. Seas tend to be thought of in the plural, as in the phrase 'the seven seas', which derived from Mesopotamian and early Greek geographers' familiarity with the Aegean, Adriatic, Mediterranean, Black, Red and Caspian seas, plus the Persian/Arabian Gulf. By medieval times, the phrase referred to the North Sea, Baltic, Atlantic, Mediterranean, Black, Red and Arabian seas, while in the centuries following the Age of Exploration the seven seas became identified globally with the Arctic, Atlantic, Indian, Pacific, Mediterranean and Caribbean seas, along with the Gulf of Mexico. This is not the whole linguistic story, of course, and the International Hydrographic Organization currently recognizes more than seventy bodies of water with the word 'sea' in their names, a list that does not include inland salt lakes such as the Aral Sea, Caspian Sea, Dead Sea or Salton Sea, or freshwater lakes such as the Sea of Galilee, none of which are classified as divisions of the World Ocean.[6]

Such geographical semantics can have far-reaching implications. In the case of the Caspian Sea, the world's largest inland body of water, the fact that it is not officially designated a sea has long complicated the political relationship between the five nations that share its coastline, along with the wealth of oil and natural gas beneath its waters. As a lake, the Caspian's resources are shared proportionally among its bordering countries – Azerbaijan, Russia, Kazakhstan, Turkmenistan and Iran – but as a sea it would be governed by the United Nations Convention on the

The freedom of the seas is in Your hands!
DON'T ~~GIVE~~ Slow UP THE SHIP!

A futuristic stroll through the Sargassum forests on the Atlantic sea floor, as imagined in 1899. The Sargasso Sea is the only designated sea with no land boundaries, and thus no shoreline. It was named after the characteristic brown Sargassum seaweed, lines of which can extend for many kilometres along the surface.

'The freedom of the seas is in Your hands!' A U.S. War Office poster, designed by Jon Whitcomb, 1943, urging maritime Americans to keep the sea lanes clear.

Law of the Sea, which would grant access to and from international waters (through Russia's Volga river and the canals that connect it to the Black Sea and Baltic Sea), along with greater access to its oil. Simply put, the states with longer coastlines have historically favoured categorizing the Caspian as a 'sea', while those with shorter coastlines have favoured categorizing it as a 'lake'. In August 2018, after twenty years of stalemate, the five Caspian nations signed the Convention on the Legal Status of the Caspian Sea, giving the body of water a unique legal status, neither lake nor sea. Each nation now has exclusive control over an area extending up to 28 km (15 nautical mi.) from its shores for mineral and energy exploration, and a further 16 km (10 mi.) for fishing: 90 per cent of the world's caviar is sourced from the Caspian Sea. The remaining area continues to be shared jointly, pending future negotiations.

'Sea' may be a small syllable, but it comes with a big editorial challenge. Should the title of this book refer to 'the sea' or 'the seas'? One generic body of water, or seven, or seventy? It is not an easy question to answer, and it is one that has long confronted translators of Jules Verne's great submarine novel, which was published in French as *Vingt mille lieues sous les mers* (1870).

Any literal translation should render the last word in the plural, since Verne's *mers* implied all seven seas, through which the novel's narrator, Professor Pierre Aronnax, voyaged in Captain Nemo's *Nautilus*. But while the first English translation, by Lewis Page Mercier (1872), rendered the 'seas' of the title as plural, most subsequent English versions reduced it to the singular, although a couple of more recent translations, including William Butcher's 1998 Oxford World's Classics edition, have reverted to the original plural.[7]

Verne himself struggled with the title of his novel, moving variously between *Journey Under the Waters*, *Twenty Thousand Leagues Under the Waters*, *Twenty-five Thousand Leagues Under the Waters*, *Twenty-five Thousand Leagues Under the Seas* and *Twenty Thousand Leagues Under the Oceans*. The cognates 'sea(s)', 'water(s)' and 'ocean(s)' have always been problematic, with the language of the sea rarely at ease when it finds itself on dry land. As will be seen over the following chapters, the sea has long been a linguistic factory, generating a wealth of slangs and argots among those whose lives have depended on it – to the bitter end, one might say, recalling an entry in Captain John Smith's *Seaman's Grammar* (1653), which explains, for the benefit of landlubberly readers, that 'a bitter end is but the turn of a cable about the bitts . . . and the bitter's end is that part of the cable doth stay within board'.[8]

Landlubbers have always dreamed of going to sea. One of the earliest English-language poems, *The Seafarer* (c. AD 750), purports to be narrated by an elderly mariner looking back on his sea-weary life, but, given the contradictory sentiments expressed in the poem, literary scholars now read it as a dialogue between an old seafarer and a young adventurer, excited by the promise of a maritime voyage:

Frontispiece to the second edition of Jules Verne's *Vingt mille lieues sous les mers* (Twenty Thousand Leagues Under the Seas).

My spirit sallies over the sea-floods wide,
Sails o'er the waves, wanders afar
To the bounds of the world and back at once,
Eagerly, longingly; the lone flyer beckons
My soul unceasingly to sail o'er the whale-path,
Over the waves of the sea.[9]

As the poem intimates, the sea is easily imagined as a place of longing, of adventure, of reinvention. For the ancient Greeks, the literary concept of *nostos* (literally, 'homecoming') invoked a heroic sea voyage from which a protagonist returns home transfigured, like Odysseus in Homer's epic, while for the Vikings, whose longships afforded swift North Sea crossings, the sea, personified as 'Ægir', was the bitter salt-road leading to wealth and acclaim. The siren call of the sea has been a creative preoccupation from Homer to the present day, and was unsettlingly personified by sculptor Antony Gormley's large-scale installation *Another Place* (1997). The piece comprises a hundred life-sized cast-iron figures that stare out over the Irish Sea from

A view of Antony Gormley's installation *Another Place*, 1997, on the sands of Crosby Beach, Merseyside. As the tide advances each day, the one hundred life-sized cast-iron figures are slowly submerged.

various points along the sands of Crosby Beach, Merseyside, where they endure a twice-daily submergence ritual at the whim of the incoming tide. The piece was originally designed in 1995 for temporary installation on the mudflats outside the port of Cuxhaven, on Germany's North Sea coast. Gormley's written proposal for the project, the most detailed he had yet submitted, outlined the scale of the idea:

> The work will occupy an area of 1.75 square kilometres, with the pieces placed between 50 and 250 metres apart along the tideline and one kilometre out towards the horizon, to which they will all be facing . . . The ones closest to the horizon will

stand on the sand, those nearer the shore being progressively buried. At high water, the sculptures that are completely visible when the tide is out will be standing up to their necks in water.[10]

The choice of location, near a busy port, allowed for a range of encounters with the figures, from a glimpse of a half-submerged iron man facing the grey immensity of the sea, to more social interactions with picnicking beachgoers as ferries and container ships pass by. The work could not be seen all at once from any of the one hundred locations, but had to be pieced together over time and distance, with the topography of the mudflats adding another layer of coastal contingency, being places where land and sea divide and merge. In the Netherlands these interspaces have generated a specific form of excursion, the *wadlopen* (mudflat walk), in which the vast mudflats of the Waddenzee are explored on foot, an exercise in geographical minimalism, in which walkers must concentrate on placing their feet safely as they proceed towards the horizon. At low tide, the mud stretches all the way to the Frisian Islands, 20 km (12 mi.) offshore, and guides are employed to steer stragglers to safety before the tide returns. When Pliny the Elder learned of the twice-daily transformation of the mudflats of the Frisians, he was unsure whether such places belonged to land or sea, their inundations blurring and confounding one of Earth's elemental boundaries, 'the line between the ordered world of *Natura* and the chaos of Ocean'.[11]

Though *Another Place* was rooted in a specific location, it has moved several times since, firstly to a fjord near Stavanger, Norway, in 1998, and then to the Belgian resort of De Panne, before finding a permanent home in northwest England in 2007, after a protracted period of negotiation with local interest groups including surfers, coastguards and conservationists concerned about public safety as well as the environmental impact of the installation, especially on migratory birds. As Gormley himself pointed out, *Another Place*'s final location, on a non-bathing beach close to Liverpool's Seaforth Dock, bears a strong

The Portuguese slave ship *Diligente* was engaged in illegal slave trading when it was seized on its way to the Bahamas by a British naval sloop in May 1838. Some six hundred enslaved people were discovered on board the vessel. Lieutenant Henry Hawker, of the Royal Navy, painted this watercolour on the spot.

relation to the Cuxhaven original, sharing as it does a similar topography, as well as a comparable legacy of maritime and colonial history marked by migration and forced displacement. Britain's last (legal) slave ship, the *Kitty's Amelia*, sailed past Crosby Beach from Liverpool harbour in July 1807, on its way to pick up its final human cargo in the Bight of Benin. Although the Act for the abolition of the slave trade had come into effect in May 1807, *Kitty's Amelia* had been granted clearance to sail at the end of April, a few days before the deadline. So when she set sail for Barbados via the slave ports of Sierra Leone, she did so within the letter of the law, but it would be the last legally sanctioned slave-trading voyage made by any English vessel.

Some fifteen years on from its final relocation, *Another Place* continues to speak to issues of displacement, especially now, in the context of the Mediterranean migrant crisis, along with the climate-driven migrations that are likely to escalate in the decades to come. In November 2017 the German newspaper *Der Tagesspiegel* published the names of the 33,293 refugees and migrants who had drowned trying to reach Europe over the previous two decades, a number that continues to rise with every year that passes. For those who attempt the journey, the sea is both a bridge and a barrier, a threshold between all manners of life and death. In 2020 more than 1,000 migrants drowned while crossing the Mediterranean, while dozens more lost their lives in the English Channel, during the final stage of what had, for many, been a long and agonizing journey.

In placing his work at the sea's edge, Gormley raises a host of questions about our relationship with maritime nature and culture, in a place where, he writes, 'time is tested by tide, architecture by the elements, and the prevalence of sky seems to question the earth's substance. In this work human life is tested against planetary time.'[12] Gormley's salt-washed figures, some now clothed in barnacles, call to mind the world's rising waters, and the drowning of vulnerable coastlines, as well as humanity's atavistic emergence from the ocean. The iron figures stare at the horizon as though struck by recognition, as though paused in some mute attempt to comprehend the immense ebbings and flowings of the sea.

1 Shorelines

The sea marks the end of things. It is where life stops and the
unknown begins.
Jonathan Raban, *Coasting* (1986)

My early childhood was spent in west Cornwall, an ancient
peninsula of granite and rain where England quietly steps off
the map and walks into the sea. On the small stony beach near
our house, my brother and I spent hours throwing pebbles into
the water, aiming at a bobbing target, or seeing who could throw
the furthest, while our father clattered around on his fibreglass
boat, the briny tang of which I can still recall, nearly fifty years
later.

I live and work in London now, and I miss the sea in ways
that are not always communicable: the sounds, the smells, the
hint of piratical adventure that flashed over the grey horizon as
we rowed around the Carrick Roads towards Falmouth and the
English Channel.

Shorelines are liminal zones, elemental boundaries of crossings and transitions that do much to shape the cognitive worlds
of coastal peoples. 'Where does the sea begin and end?' is a
deceptively simple question, for while it is easy to define the
shoreline – 'the line where shore and water meet', in the words
of the *Oxford English Dictionary* – it is more difficult to find it,
for a shoreline is a mutable space, its contours erased and rewrit-
ten twice a day by tidal variations. To watch an incoming sea
encroach onto the land, then enact its slow reversal six hours
later, is both a local and a cosmic experience, the demarcations
of high and low tide merely the visible traces of planetary processes that are otherwise too big to see, with the Sun and the
Moon exerting powerful gravitational drag on the waters of the

rotating Earth. The Moon pulls our seas up towards it, creating a bulge in the oceans, which is then sent sweeping around the rotating planet by the near-equal force of inertia. This ever-moving ripple is experienced at the surface as the twice-daily tides, the ebbings and flowings of which have been objects of fear and fascination since before the expeditions of Alexander the Great (fourth century BC), whose generals, accustomed to the virtually tideless Mediterranean, were reportedly alarmed by the rise and fall of the waters of the Indian Ocean.

My older brother and I playing on the shoreline, Feock, Cornwall, c. 1970.

Watching sea waves advance and break is, as Jonathan Raban observes, 'a pastime designed to induce reflective melancholy', and shorelines have long associations with pensiveness and brooding, from Shakespeare's Sonnet 60 ('Like as the waves make towards the pebbl'd shore, / So do our minutes hasten to their end') to Matthew Arnold's 'Dover Beach' (1867), in which the sound of the advancing sea calls to mind the painful ebb and flow of human history:

Where the sea meets the moon-blanched land,
Listen! you hear the grating roar
Of pebbles which the waves draw back, and fling,

At their return, up the high strand,
Begin, and cease, and then again begin,
With tremulous cadence slow, and bring
The eternal note of sadness in.[1]

The poem, which was begun on honeymoon in the summer of 1851, enacts a movement from pleasure to pessimism, the latter drawing in like an unsettling tide that echoes 'the turbid ebb and flow of human misery'. The poem's central motif, the withdrawing 'Sea of Faith', is a variant of 'the unplumbed, salt, estranging sea' that was the defining emblem of Arnold's earlier 'Marguerite' poems, which explored similarly pessimistic territory.[2] Arnold once described himself as 'one who looks upon water as the mediator between the inanimate and man', and images of seas and rivers featured prominently in his work, symbolizing, in the words of the Arnold scholar Miriam Allott, 'the hidden currents of the buried self', whose true nature, once known and understood, would rejuvenate an unhappy life.[3]

Many of the circumstances of the setting and imagery of 'Dover Beach' would be borrowed by Ian McEwan for his honeymoon novella, *On Chesil Beach* (2007), in which the incoming waves, 'whose steady motion of advance and withdrawal made sounds of gentle thunder, then sudden hissing

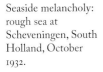

Seaside melancholy: rough sea at Scheveningen, South Holland, October 1932.

against the pebbles', prefigure the sexual catastrophe to come.[4] Later in the story, after a humiliating episode, Florence runs onto the beach, followed by Edward; they argue, before Edward is left on his own to brood at the water's edge, letting the waves wash over his shoes. As McEwan's unhappy couple discovers, shorelines are places of stark exposure, where modernity recedes and the elements regain their ancient power, exemplified by the way that centuries of storm waves had sifted and graded the pebbles along the beach, depositing them in neat size order from east to west: 'The legend was that local fishermen landing at night knew exactly where they were by the grade of shingle. Florence had suggested they might see for themselves by comparing handfuls gathered a mile apart', but the tragic, Arnoldian misunderstandings that unfold over the rest of the novel conspire to thwart their littoral investigations, before blighting the rest of their lives.[5]

Topography

The 180 billion pebbles that form Chesil Beach have indeed been sorted by the sea into size order, with the smallest found at West Bay and the largest at the Isle of Portland, 29 km (18 mi.) to the east. Chesil Beach (its name derives from the Old English *cisel*, 'shingle') is an extensive barrier beach that formed when pebbles were washed up en masse at the end of the last ice age, more than 5,000 years ago. Behind the 200-metre-wide (660 ft) strip of beach lies the Fleet, a shallow tidal lagoon cut off from the English Channel, and rich with fossils, as are many other sites along southwest England's Jurassic Coast.

The sea-sculpted complexity of Earth's coastlines, with their inlets, river mouths, firths, fjords, promontories, beaches, tombolos and isthmuses, is the product of what Robert Macfarlane has called 'the dialogue between solid and liquid', the relative hardness, softness, porosity or impermeability of coastal materials determining how the actions of seawater will shape and refine them over time.[6] If there is poetry in these languages of coastal topography, it has arisen in response to the intricacy of its geophysical arrangements, and the classificatory challenge of a landscape in perpetual flux. Beaches have their 'breaker zones', 'swash zones', 'surf zones', 'transition zones', 'backshores', 'foreshores' and the like, while a glance at the 'Coastlands' section of Macfarlane's *Landmarks* (2015), his celebration of the global topographical lexicon, confirms how many words reflect the changeability of the littoral landscape, from Shetland's 'ar'ris' (the last weak movements of a tide before still water) to the Cornish 'zawn' (a vertical fissure or cave cut into a cliff by waves), via 'gunk-hole' (a narrow channel rendered too dangerous to navigate by the current); 'oyce' (a lagoon formed where a bar of shingle has been thrown up across the head of a bay); 'tombolo' (a ridge of sand built by waves that connects an island to the

The complexity of the shoreline, illustrated by the intricate patterns of braided streams, coloured by marine algae, on the coast of Admiralty Island, Alaska.

mainland); or 'vaddel' (a gulf that fills and empties with the flowing and ebbing of the sea).[7]

As such language attests, near-shore topography is a world apart from the open sea, calling for different forms of experience and understanding, as well as a different vocabulary. Naval captain Greenville Collins, in his *Great Britain's Coasting Pilot* (1693), the first official survey of the British coastline, drew a memorable distinction between the skills needed for navigation at sea and those needed in inshore waters: 'The Marriner having left the vast Ocean, and brought his Ship into Soundings near the Land, amongst Tides or Streams, his Art now must be laid aside, and Pilottage taken in hand, the nearer the Land the greater the Danger, therefore your care ought to be more.'[8]

The risk to shipping increases as land draws near, the shoaling effect of the rising sea floor affecting the complex interactions of waves, currents, seabed and shore. As a wave enters shallow water, its height increases while its wavelength decreases; as the wave steepness (the ratio of wave height to wavelength) increases, so the wave becomes less stable. The movement of a wave up a beach is known as the 'swash', with its reverse movement known as the 'backwash', the relative strength and stability of these movements determining how dangerous a wave might be to approaching craft, whether the wave is a 'spilling wave' (a softer and more consistent wave that breaks gradually as it approaches shore); a 'dumping wave' (a low-tide wave that breaks powerfully in shallow water); or a 'surging wave' (a non-breaking wave that retains its power and momentum). All three wave types are made of water on the move, unlike the waves of oscillation encountered out at sea, in which energy, not water, is transferred from crest to crest. These beach waves, known as waves of translation, act differently from sea waves, as they race up the beach, riding over one another in chaotic, hazardous patterns that can be difficult – and dangerous – to navigate.

In Stephen Crane's autobiographical story 'The Open Boat' (1897), a foundered ship's lifeboat off the coast of Florida turns back from land when the surging waves at the surf zone prove too dangerous to approach:

Wind-driven wave breaking in the surf zone, blown in from a storm over the Atlantic.

The billows that came at this time were more formidable. They seemed always just about to break and roll over the little boat in a turmoil of foam. There was a preparatory and long growl in the speech of them. No mind unused to the sea would have concluded that the dinghy could ascend these sheer heights in time. The shore was still afar. The oiler was a wily surfman. 'Boys,' he said swiftly, 'she won't live three minutes more, and we're too far out to swim. Shall I take her to sea again, Captain?'

'Yes; go ahead!' said the captain.

This oiler, by a series of quick miracles and fast and steady oarsmanship, turned the boat in the middle of the surf and took her safely to sea again.[9]

Waves out at sea can be ridden more safely than surging and breaking waves near land, wave energy being transferred across open water rather than dissipated at the surf zone, where shoaling waves approaching a shoreline are bent, or refracted, focusing wave energy. As waves enter shallow water, interaction with the sea floor alters the waves, decreasing their speed while increasing their height. These shallow-water waves become so high and unstable that they break to create surf, during which chaotic process they erode and transport sediment, altering the topography of the sea floor, which, in turn, affects the character of the following wave. In Crane's story, the lifeboat is later capsized by a large unstable wave during a second attempt to beach, and the four occupants are pitched into the cold January sea, where they must swim against the rollers to reach the shore.

As Greenville Collins observed in his *Coasting Pilot*, 'it sometimes happens, and that too frequently, that when Ships which have made long and dangerous Voyages, and are come home richly Laden, have been shipwrack't on their Native Coast', and it was for just such situations that the *Pilot* was produced, offering detailed guidance for sailing into every port or inlet in the country.[10] The entry on Falmouth and the Carrick Roads, the scene of my childhood rowing expeditions, gives a sense of the book's detailed guidance:

You may Sail in and out of either side of Falmouth-Rock; the east side is the best. Being past the Rock, and that you would Anchor in Carreck-Road, which is the Place where great Ships ride, you may Sail up in the fair way, keeping your Lead; for there is a narrow deep Channel which hath eighteen Fathom water all the way up to Carreck-Road; you may borrow on St Mawes side in five and six Fathom water: the west side is shoal.[11]

Collins's description still holds true today, more than three hundred years on, the main channel through the Carrick Roads remaining 34 m (18 fathoms) deep, although the 'great Ships' have largely been replaced by small-scale leisure craft, piloted by weekend enthusiasts rather than Collins's 'Soveraigns of the Seas'.

In seventeenth-century Cornwall, of course, seashore topography was not the only perceived danger to shipping. The Cornish coast was notorious for 'wreckers', figures of nightmare who made their living luring ships onto rocks before helping themselves to the drowned sailors' cargo. Daphne du Maurier's *Jamaica Inn* (1936) and Winston Graham's Poldark novels did much to reinforce the association of Cornwall with 'false lights' wrecking, the historical extent of which remains a source of emotive debate. After all, the notorious Cornish prayer, 'Oh Lord, let us pray for all on the sea; but if there's got to be wrecks, please send them to we', invokes scavenging rather than deliberate wrecking, for which there were notably few convictions, with only one person in Cornwall ever executed for the crime: an eighty-year-old farmer named William Pearce, who was hanged in 1767 for his part in plundering a ship 'whose Cables were cut, and she made a Wreck of as soon as the sailors had left her'.[12] The only violence that night was perpetrated against the vessel, from which the unfortunate Pearce was found to have stolen only a length of rope.

Tales of 'false lights' wrecking arose on coastlines around the world during the age of sail, but there is little evidence for it ever having been a genuine practice. In hundreds of admiralty court cases heard in the busy ports of Florida, no captain of a wrecked ship ever claimed to have been led astray by a false light, while a nineteenth-century Bahamian wrecker, when asked if he and his crewmates made beacons on shore or showed their lights to lure ships, is reported to have said, 'no, no [laughing]; we always put them out for a better chance by night', since mariners would usually interpret a light as indicating land, and so avoid it, particularly if it was unexpected.[13] In any case, hand-held oil lanterns could hardly be seen from out at sea, unless fitted with mirrors or lenses, and mounted at an elevated height, in other words, if they were built as replica lighthouses. Lighthouses had been familiar coastal landmarks since ancient times, though some of the earliest functioned more as entrance markers to ports than as warnings against rocks and reefs. The oldest lighthouse still in use is the 55-metre (180 ft) Torre de Hércules (Tower of Hercules), a second-century Roman lighthouse built on a peninsula in northwest Spain, to a basic design that has hardly changed over the intervening centuries, despite innovations in power supply and automation. But with ports and lighthouses clearly marked on every mariner's chart, it seems unlikely that a hand-swung lantern was ever mistaken for one, unless a vessel was already foundering in a storm, driven on towards the fateful shore.

If wrecking was a long-established feature of the coastal imaginary, this had been fuelled in part by legal definitions enshrined in various acts, such as the British Wreck Act of 1275 (contained within the Statute of Westminster), which ruled that

South Pier Lighthouse, Penzance, Cornwall, built in the 1850s to replace an earlier structure that was washed away in a storm.

The Tower of Hercules, on the coast of Galicia, northern Spain. Built by the Romans in the 2nd century AD, and restored in 1791, it is the oldest working lighthouse in the world.

if anyone survived a shipwreck, then the ship's cargo could not be considered legal salvage. The Act's unfortunate wording – 'Concerning Wrecks of the Sea, it is agreed, that where a Man, a Dog, or a Cat escape quick [alive] out of the Ship, that such Ship nor Barge, nor any Thing within them, shall be adjudged Wreck' – appeared to offer an incentive to plunderers to ensure that no one, not even a cat, made it alive to shore.[14] Although the clause had been intended to discourage wreckers and salvors from seizing or damaging seaworthy craft, it read, in effect, as a murderers' charter, and whether or not coastal robbers ever interpreted this law to the letter – in the manner of *Jamaica Inn*'s

murderous Joss Merlyn – the 'man nor beast' rule was familiar
enough to fuel a fearsome wrecking mythology, of the sort
expressed by Daniel Defoe in the course of his *Tour thro' the
Whole Island of Great Britain* (1724–7), in which he maligned the
inhabitants of the Scilly Isles as

> a fierce and ravenous people, for they are so greedy, and eager
> for prey, that they are charged with strange, bloody and cruel
> dealings ... especially with poor distressed seamen when
> they come on shore by force of a tempest, and seek help for
> their lives, and where they find the rocks themselves not
> more merciless than the people who range about them for
> their prey.[15]

The man nor beast rule was finally repealed in the 1770s, with
historians continuing to argue over whether it actually incited
violence against mariners, or was merely a badly worded law with
unintended myth-making consequences.

Scavenging for wrecked cargo remains illegal, in spite of
what beachcombers today might think. In January 2007 the container ship MSC *Napoli* ran aground off the coast of Devon,
spilling more than a hundred shipping containers onto Branscombe Beach. News footage of locals plundering the washed-up
flotsam was shown around the world, with the scavenged items
including motorcycles, nappies, perfume and car parts. After tol-
erating a brief free-for-all, the police closed the beach and
announced that they would employ powers not used for a century to force the return of the taken goods. The act of collecting
washed-up cargo – 'aggravated beachcombing', as it is sometimes
known – is enmeshed in complex legal and linguistic frame-
works; but in spite of nearly a thousand years of legislation to the
contrary, the false belief that everyone has the right to help themselves to objects found on a beach persists. This is partly due to a
perception that the shoreline itself is public property, or rather
that it belongs to no one in particular, and so anything found
there is fair game. But salvage laws state that just because something has washed ashore does not mean it no longer has an

Horseback
shrimping contest,
Oostduinkerke,
Belgium, 1925.

owner. Salvors may be able to claim a reward based on the value
of rescued property, but the law requires all finds to be registered
with the Receiver of Wreck, a post created under the auspices of
the Merchant Shipping Act 1995, a wide-ranging piece of legislation that consolidated and updated an array of British maritime
legislation, while providing a concise taxonomy of the forms of
wreccum maris, the spoils of the sea:

> *Flotsam*: goods lost from a ship which has sunk or otherwise
> perished which are recoverable because they have floated.
> *Jetsam*: goods cast overboard (jettisoned) in order to lighten
> a vessel which is in danger of sinking, even if they ultimately
> perish.

Derelict: property which has been abandoned and deserted at sea by those who were in charge without any hope of recovering it. This includes both vessels and cargo.
Lagan (or ligan): goods cast overboard from a wrecked ship, buoyed or moored so that they can be recovered later.

The legal principle governing all four categories affirms that wrecked or salvaged goods continue to belong to their original owners, and that any wreckage remaining unclaimed after a year belongs to the Crown. But the law is one thing, practice is another, and beachcombing, whether 'aggravated' or not, remains widely viewed as a benign activity sanctified by coastal tradition; as many of the salvors on Branscombe Beach and elsewhere have stated, 'They won't stop us doing it – it's our culture. It's in our blood.'[16]

The concept of 'beachcombing' has its roots in ethnography, referring specifically to foraging practices on the islands of the South Pacific. The word was coined by Richard Henry Dana Jr

Beachcomber with horse and wagon gathering wreckage on a Dutch beach, 1932.

in 1840, in reference to a cohort of European sailors, mostly disgruntled whalers or escaped convicts, who had settled in the Pacific islands to make their living by foraging, pearl fishing 'and often by less reputable means', as the *Oxford English Dictionary* phrases it. During the sixteenth and seventeenth centuries, a small number of European castaways ended up living on islands on the main trans-Pacific sailing routes (such as Magellan's cabin boy, Gonçalo de Vigo, who deserted in the Marianas in 1521, and spent four years with the Chamorro people before being picked up by another explorer), but the beachcombing era began in earnest in the late eighteenth century when commercial shipping radiated out from New South Wales and North America: Hawaii was a favourite haunt, along with Tahiti, Tonga, Fiji and Samoa. By the 1850s there were several thousand such characters scattered over the islands of Polynesia and Micronesia, as well as on New Zealand, where they were known as 'pakeha-Maoris'.[17]

The idea of the Crusoe-esque beachcomber became a popular fixture of the European imagination, with Joseph Conrad's youthful seagoer 'Lord Jim' depicted imagining himself 'as a lonely castaway, barefooted and half-naked, walking on uncovered reefs in search of shell-fish to stave off starvation'.[18] But the image of philosophical castaways living carefree lives under swaying palm trees did not reflect the harsh reality of what was often a precarious and lonely existence, and the majority of European beachcombers stayed only a few months or years before making their way back home, or at least onto another passing ship. As one such disillusioned seaman explained upon his return from Samoa in the 1850s, the boredom of his island paradise became too much to bear:

> Here, I at first thought, my dreams of island felicity were to be realized . . . [but] the gloss of novelty wore off in a few weeks, and disclosed the bareness and poverty of savage life, even in its most inviting forms. I grew weary of lying all day long in the shade, or lounging on the mats of the great house, or bathing in the bright waters.[19]

Clearly, paradise has its limits, as does the dream of endless leisure, and as this opening chapter has sought to show, shorelines have always been complex spaces of arrival and departure, of embarkation and emigration, of longing and change, proving more conducive to short-lived bouts of introspection than to long-term inactivity, which is why beachcombers in the end have always grown tired of their littoral lives. And perhaps, as will be seen in the following section, it is why the traditional seaside holiday has rarely lasted longer than a fortnight.

Modern beachcombers working at low tide along the shoreline of Viti Levu, Fiji.

The seaside

According to historian Alain Corbin, the seaside holiday was 'invented' more or less single-handedly by a Reverend William Clarke and his wife, who spent a month in the summer of 1736 'sunning ourselves upon the beach at Brighthelmstone', a declining fishing town soon to be reborn as a seaside resort.[20] As Clarke observed in a letter to a friend:

> The place is really pleasant; I have seen nothing in its way that outdoes it. Such a tract of sea; such regions of corn . . . but then the mischief is, that we have little conversation besides the *clamor nauticus*, which is here a sort of treble to the plashing of the waves against the cliffs. My morning business is bathing in the sea, and then buying fish; the evening is riding out for air, viewing the remains of old Saxon camps, and counting the ships in the road, and the boats that are trawling.[21]

Although there was nothing new about a seaside sojourn, the Clarkes' innovation had been to transpose the conventions of the inland spa holiday to a seaside setting, with the addition of immersive sea bathing, as practised by local working people 'for a combination of therapeutic, prophylactic, educational, festive and hedonistic purposes'.[22] If Brighthelmstone had lacked any kind of holiday infrastructure in 1736, thirty years later the town had transformed into Britain's most popular and fashionable resort, with a new name, Brighton, that became widely known across the English-speaking world as a shorthand for seaside glamour.

The transformation of coastal towns into seaside resorts was driven partly by technological change – smaller settlements such as Brighthelmstone and Worthing could no longer compete with larger fishing towns with capacious trawler basins and deep-water docks – and partly by a growing awareness of the medical benefits of sea air and salt water. The publication, in 1750, of Brighton-based doctor Richard Russell's *Dissertation on the Use of Sea Water in the Diseases of the Glands* (he pronounced seawater to be 'superior to the waters of inland spas') has long been credited with kick-starting seaside mania in Britain. A plaque on the site of his practice in Brighton reads: 'if you seek his monument, look around', while the French historian Jules Michelet went as far as describing Russell as 'the inventor of the

DR. RICHARD RUSSELL
F. R. S. (1687-1759)
AUTHOR OF A DISSERTATION
CONCERNING THE USE OF SEA
WATER IN DISEASES OF THE
GLANDS (1750). FOUNDER OF
BRIGHTON AS A
BATHING RESORT.
BORN AND PRACTISED
MEDICINE IN THIS
HOUSE.   19   88

Plaque commemorating Dr Richard Russell, promoter of sea bathing, at his birthplace in Lewes, East Sussex.

sea'.[23] For Russell, drinking the sea was as important as bathing in it – among his daily recommendations were 'cold bathing in the Sea, drinking every Morning enough Sea Water to procure two or three Stools a Day, immediately upon coming out of the Sea' – but over time it would be the restorative properties of sea air and sunshine, rather than the consumption of salt water, that did most to popularize the seaside holiday, especially after King George III took to the sea at Weymouth in the summer of 1789, while a chamber orchestra played the national anthem from inside a neighbouring bathing machine.[24]

The Quaker physician John Coakley Lettsom, a pioneer of open-air treatment for respiratory complaints, founded Margate's Royal Sea Bathing Hospital in 1791, a large compound built around a solarium that bathed the patients in sea air and sunlight. By the turn of the nineteenth century, just about any illness could be claimed to be cured by a week at the seaside, with daily immersion in 'nature's richly saturated compound' made possible by the invention of the bathing machine, 'a curious contrivance of Wooden Houses moveable on wheels', as one observer described them in 1745.[25] Tobias Smollett's autobiographical novel, *The Expedition of Humphry Clinker* (1771), features a memorable account of their use at Scarborough:

Image to yourself a small, snug, wooden chamber, fixed upon
a wheel-carriage, having a door at each end, and on each side
a little window above, a bench below – The bather, ascending
into this apartment by wooden steps, shuts himself in, and
begins to undress, while the attendant yokes a horse to the
end next the sea, and draws the carriage forwards, till the
surface of the water is on a level with the floor of the dressing-
room, then he moves and fixes the horse to the other end
– The person within being stripped, opens the door to the
sea-ward, where he finds the guide ready, and plunges head-
long into the water – After having bathed, he re-ascends
into the apartment, by the steps which had been shifted for
that purpose, and puts on his clothes at his leisure, while the
carriage is drawn back again upon the dry land; so that he
has nothing further to do, but to open the door, and come
down as he went up.[26]

As Smollett's male narrator goes on to observe, the machines
could only be used during certain tidal phases, which varied
from day to day, but 'for my part, I love swimming as an exercise,
and can enjoy it at all times of the tide, without the formality of
an apparatus . . . you cannot conceive what a flow of spirits it
gives, and how it braces every sinew of the human frame.'[27]

The passage offers a sense of how, even on an eighteenth-
century beach, inhibitions could be cast aside with the clothes.
The framing of the beach as a debatable land where codes of
propriety are temporarily suspended was already under way,
although a degree of anxiety was also expressed over the pros-
pect of people of different classes and sexes mixing at the
shoreline in various states of undress. These anxieties connected
with longer-established forms of disquiet over the morality of
ports, and the sexual licence enjoyed at sea, and Brighton's reputation as a gentrified resort competed with its growing reputation
as a locus of extra-marital sex. By the 1850s there were nearly a
hundred brothels in the town, and its association with the 'dirty
weekend' was well established. Long before the advent of the
sexualized seaside postcard in the 1920s, or of suggestive music

hall songs such as 'You Can Do a Lot of Things at the Seaside (That You Can't Do In Town)' cartoonists reflected the often unsettling encounters between the fully and the partially clothed within the permissive location of a seaside town. In Thomas Rowlandson's *Summer Amusement at Margate, or a Peep at the Mermaids* (1813), groups of clothed men gather near the shoreline to gaze at a party of naked female swimmers in the water. A clothed woman arrives to upbraid her husband with a parasol, anticipating the sexual-domestic comedy of later postcard production by the likes of Donald McGill.

As these descriptions make clear, sea bathing was a separate activity from swimming; in fact, remarkably few of the eighteenth- or nineteenth-century tourists who ventured nervously into the waves would have been able to swim, and many would have been horrified at the idea. Those who did swim in the sea tended to do so vigorously, for exercise rather than for pleasure, as recommended in a 'seaside manual for invalids' published in 1841:

Thomas Rowlandson, *Summer Amusement at Margate, or a Peep at the Mermaids*, 1813, hand-coloured etching. Clothed men gaze at a party of female sea bathers.

SUMMER AMUSEMENT AT MARGATE, OR A PEEP AT THE MERMAIDS.

The common way of swimming is with the face toward the water. In this situation the body lies extended, prone, upon the fluid – with the mass of muscles on the back in a state of powerful and permanent contraction, so as to fix the hips and spine, and keep back the head. The muscles of the limbs are in somewhat rapid action, propelling the body through the water . . . In the graceful kind of natation, called walking in the water, the muscles of the chest are principally called into exercise; and the organs which it contains thereby acquire volume and force; but requiring considerable effort, it cannot be kept up long.[28]

It was only after the introduction of more user-friendly strokes, such as the front crawl, that swimming in the sea became a widespread activity. In August 1875 Matthew Webb, a former Cunard Line captain, became the first recorded person to swim the English Channel, although it was far from a straightforward journey; Webb was attacked by jellyfish, and pulled wildly off course by strong currents. At 64 km (40 mi.), his circuitous route was nearly double the distance of the shortest direct crossing, 33 km (20 mi.) from Dover to Calais, but nonetheless the bruised and porpoise-oil smothered Webb set a time that would not be bettered for another fifty years.

By the mid-nineteenth century, all the elements that made up a seaside resort had been established, from piers and promenades from which to enjoy sea views and breathe healthful sea air, to winter gardens, seaside shelters, souvenir shops, tearooms and aquaria that provided rest and amusement away from the beach itself. Rows of purpose-built hotels and guest houses capitalized on the increasing demand for views of the sea, with 'Sea View' itself becoming the archetypal name for a seaside guest house or bungalow.[29] In fact visual proximity to the sea became one of the most significant aspects of a seaside holiday ('I can see the sea!'), and the design and positioning of balconied hotels on coastlines across the world continues to reflect a nineteenth-century fixation on rooms with a view of the waves.

In Jane Austen's *Persuasion* (1817) Anne Elliot and Henrietta Musgrove visit the sands at Lyme Regis before breakfast one morning in order to watch the incoming tide, 'which a fine south-easterly breeze was bringing in with all the grandeur which so flat a shore admitted. They praised the morning; gloried in the sea; sympathised in the delight of the fresh-feeling breeze – and were silent.'[30] Such seaside sublimity contrasts with the bustle of the coastal village in Austen's last, unfinished, novel *Sanditon* (1817), which was depicted in the unfortunate throes of being developed into a 'fashionable bathing place' by Tom Parker, Austen's tireless speculator, who boasted of 'the finest, purest Sea Breeze on the Coast – acknowledged to be so – Excellent Bathing – fine hard Sand – Deep Water ten yards from the Shore – no Mud – no Weeds – no slimey rocks – Never was there a place more palpably designed by Nature for the resort of the Invalid', while his contention that 'no person could be really in a state of secure and permanent health without spending at least six weeks by the sea every year' recalled John Coakley Lettsom's pioneering ideas on the healthfulness of maritime air:

> The Sea Air and Sea Bathing together were nearly infallible, one or the other of them being a match for every Disorder of the Stomach, the Lungs or the Blood; They were anti-spasmodic, anti-pulmonary, anti-septic, anti-bilious and anti-rheumatic. Nobody could catch cold by the Sea, Nobody wanted appetite by the Sea, Nobody wanted Spirits. Nobody wanted Strength – They were healing, softing, relaxing – fortifying and bracing – seemingly just as was wanted – sometimes one, sometimes the other. If the Sea breeze failed, the Sea-Bath was the certain corrective; and where Bathing disagreed, the Sea Breeze alone was evidently designed by Nature for the cure.[31]

Aspects of Sanditon are likely to have been based on Worthing, Sussex, a former fishing hamlet, where Austen holidayed with her family in 1805 while it was still being developed

French railway poster advertising the beach resort of Trouville, *c.* 1920s. The image emphasizes the pleasures of waves and sea views ('Hôtel Bellevue').

into a commercial resort by the entrepreneur Edward Ogle (or 'a village of indigenous fishers and farmers perverted into a place for the rootless and self-indulgent' in Janet Todd's well-wrought phrase).[32] It was there that Austen first heard of Nelson's victory at Trafalgar – which was particularly welcome news, as her brother Francis was a Royal Navy officer on active service in the Mediterranean – and twelve years later she named Tom Parker's Sanditon residence 'Trafalgar House' in honour of the association.

Parker's high-toned plans for Sanditon represented a move away from the rowdyism of holiday resorts such as Brighton and Margate, towards the more genteel retirement resort typified by Eastbourne or Bognor, the 1807 guidebook to the latter claiming that 'there was no spot on the coast of England better calculated for the two-fold purpose of sea-bathing and retirement'.[33] Bognor had been purpose-built by the entrepreneur Sir Richard Hotham (who had christened it Hothamton, a name that never caught on) in the hope of attracting high-class clients and long-stay retirees. Like Sanditon, it had a hotel, a

Edwardian-era children enjoying the seaside. Illustrated postcard by Cicely Mary Barker, c. 1925.

subscription library, milliners' shops and bathing machines, and was commended by the *Evangelical Magazine* in 1816 as a place where its readers could enjoy sea air 'without fear of dissipations' (though the coming of the railway in 1864 reversed its genteel fortunes, cemented by the building of a Butlins holiday camp a century later).[34]

The romantic idea of retiring to the coast gained purchase over the nineteenth and twentieth centuries, with seaside towns transfigured into sites of end-of-life care as well as of leisure, health and restoration. For the young Edmund Gosse, who moved with his father to the Devon coast following the death of his mother in 1857, the seashore became a kind of studious sanctuary, an outdoor laboratory where the secrets of nature might reveal themselves to him. 'No other form of natural scenery than the sea had any effect upon me at all', he recalled; 'it was the Sea, always the sea, nothing but the sea . . . My greatest desire was to walk out over the sea as far as I could, and then lie flat on it, face downwards, and peer into the depths.'[35] As he recalled in his childhood memoir, *Father and Son* (1907), he even drank seawater in an effort to get closer to the medium, hoping that a form of natural magic would enable him to lie on the water as though on a glass-bottomed boat, to feast his eyes on the in-situ aquarium below.

Gosse's childhood passion for the sea was matched by that of the poet Algernon Swinburne, whose biographer and champion Gosse later became. In his *Life* of Swinburne (1917), Gosse observed that of all the pleasures in the young poet's life 'the sea took the foremost place', and he cited Swinburne's own speculation that the sea '*must* have been in my blood before I was born'.[36] The opening chapter of Swinburne's unfinished autobiographical novel *Lesbia Brandon* (first published in 1952, long after its author's death) elaborates on this idea, describing the moment when the eleven-year-old protagonist, Herbert Seyton, is struck at first sight by what would become – as it had been for Swinburne himself – a strong somatic identification with the sea:

The colours and savours of the sea seemed to pass in at his eyes and mouth; all his nerves desired the divine touch of it, all his soul saluted it through the senses.

'What on earth is the matter with him?' said Lord Wariston.

'Nothing on earth,' said his sister; 'it's the sea.'[37]

Over the following months, 'the soul of the sea entered him', and he only felt truly alive when he was in or near the water: 'all the sounds of the sea rang through him, all its airs and lights breathed and shone upon him: he felt land-sick when out of the sea's sight, and twice alive when hard by it'.[38] Swinburne's self-diagnosed land-sickness – a fantastical corollary to seasickness – took the form of an overwhelming yearning for the sea that gripped both mind and body whenever the shore was out of sight, in an extreme manifestation of the kind of maritime spell invoked in the opening chapter of Herman Melville's *Moby-Dick* (1851), with its vision of 'the crowds of water-gazers' drawn to the New York harbour-side on Sunday afternoons:

> Posted like silent sentinels all around the town, stand thousands upon thousands of mortal men fixed in ocean reveries. Some leaning against the spiles; some seated upon the pierheads; some looking over the bulwarks of ships from China; some high aloft in the rigging, as if striving to get a still better seaward peep. But these are all landsmen . . . look! here come more crowds, pacing straight for the water, and seemingly bound for a dive. Strange! Nothing will content them but the extremest limit of the land; loitering under the shady lee of yonder warehouses will not suffice. No. They must get just as nigh the water as they possibly can without falling in.[39]

Melville's 'landsmen' draw as close to the water as they can, but stop short of physically entering it, pausing at 'the extremest limit of the land'. Only their seaward gazes cross the liminal borderline to which they find themselves pulled as though under a spell, in an image of enchantment that recurs across centuries of maritime writing, from Homer's *Odyssey* to Conrad's *Lord Jim*. The Victorian art critic John Ruskin recalled how, as a child, he was permitted to look at the sea from the safety of the promenade but not to venture into it: 'I was not allowed to row, far less to sail, nor to walk near the harbour alone; so that I learned nothing of shipping or anything else worth learning, but spent four or five hours every day in simply staring and wondering at the sea.'[40] Like Melville's collective 'ocean reveries', Ruskin's solitary 'staring and wondering' constituted a comparable visual border crossing, a means of stepping over the saltwater threshold that separated the civilized world of the seaside promenade from the world of adventure and risk. Little wonder Ruskin went on to become preoccupied by the sea, writing a rhapsodic monograph on the harbours of England, and devoting a long section of volume 1 of *Modern Painters* (1843) to 'The Truth of Water', in which he likened the task of comprehending the 'wild, various, fantastic, tameless unity of the sea' to trying to paint a soul.[41] For

Holidaymakers cooling off in the sea at Scheveningen, South Holland, July 1952.

Ruskin, as it would be for Jung, the sea – 'to all human minds the best emblem of unwearied, unconquerable power' – was essentially unfathomable, an idea that will recur throughout this book, and particularly in the following chapter, which looks at the history of ocean science, or rather, at the history of humanity's determined, if quixotic, attempts to fathom the literally unfathomable.

2 The Science of the Sea

I wonder at the sea itself, that vast Leviathan, rolled round the
earth, smiling in its sleep, waked into fury, fathomless, boundless,
a huge world of water-drops. Whence is it, whither goes it, is it of
eternity or of nothing?

William Hazlitt, *Notes of a Journey through France and Italy* (1826)

In January 1992 a dozen steel shipping containers were washed
from the deck of the cargo vessel *Ever Laurel* during a mid-
Pacific storm off Hawaii. One of the containers was filled with
a consignment of over 28,000 'Friendly Floatees', Chinese-made
plastic bathroom toys in the shape of ducks, frogs, turtles and
beavers. The fated container broke apart in the sea, setting the
Floatees adrift in the Pacific, from where they embarked on an
extraordinary global journey that is still in train, more than a
quarter of a century on from their release. The wanderings of
this colourful plastic armada have been charted in forensic detail
by oceanographers, who have welcomed the opportunity of a
larger than usual sample size of 'drift markers' with which to
track long-term movements of ocean currents and gyres around
the world.

The first Floatees to make landfall washed up on the Alas-
kan coast some ten months after the accident, having drifted
some 3,200 km (2,000 mi.) from their release point. Over the
years, more of the distinctive toys have been found on Pacific
coastlines as far away as Australia and South America, while
others headed north towards the Arctic, where they were frozen
into winter sea ice. A number of ducks turned up on British and
Irish shores in 2007, fifteen years and 27,000 km (17,000 mi.)
distant from their Pacific spillage point, having been encased in
Arctic ice for several years while en route to the Atlantic. The
scientists who studied them were surprised by the toys' longevity,
having assumed that they would quickly break down under the

abrasive actions of sun, sea and ice. But the toys had been built to survive 52 dishwasher cycles, as well as the ruthless attentions of toddlers, and proved robust enough to withstand whatever the oceans threw at them during their long careers afloat, including bites from curious marine animals. According to marine scientist Curtis Ebbesmeyer, who has spent his career tracking ocean currents, the toys proved an 'oceanographic bonanza' that offered a new way of understanding the periodic relationships of all the seas' currents and gyres, and the way that these fit together like gears in a planetary clock.[1]

The idea that the world's oceans might form a single circulating system first gained purchase in the mid-nineteenth century, at the dawn of oceanography, and it remains an area of intense scientific interest that is greatly assisted by the regular accidental losses that are a feature of maritime trade. Around 1,000 shipping containers are lost at sea every year, with their spilled contents offering scaled-up versions of a tried and tested research method that had been introduced at the outset of ocean science. The u.s. naval officer and natural scientist Matthew Fontaine Maury, writing in the 1850s, described labelled bottles thrown overboard as 'mute little navigators', and cited the work of Alexander Bridport Becher, a British naval officer who in the 1840s and 1850s had produced 'Bottle Charts' of the Atlantic Ocean, showing the tracks of nearly two hundred numbered bottles that had been deliberately dropped from ships as a means of studying ocean currents.[2]

Becher's research coincided with a worldwide fad for sending messages in bottles that had been boosted by Edgar Allan Poe's unsettling story 'MS. Found in a Bottle', which had been published in 1833 to great acclaim. In the story, a mariner on an ill-fated vessel lost in icy northern waters writes a desperate final message: 'at the last moment I will enclose the MS. in a bottle, and cast it within the sea'.[3] The ship drifts through gathering sea ice for days on end, but all hope vanishes when a whirlpool appears – 'a gigantic amphitheatre, the summit of whose walls is lost in the darkness' – and as the vessel begins to circle to its doom, the message is completed and flung overboard, 'amid a

More than 1,000 shipping containers are lost at sea every year, constituting a serious hazard to shipping and marine life (although some lost containers have been discovered acting as artificial reefs on the sea floor).

roaring, and bellowing, and thundering of ocean and of tempest, the ship is quivering – oh God! and – going down!'[4]

The story spoke eloquently to an era preoccupied by the new technologies of long-distance communication. The first electric telegraphs had appeared in 1832, and by the time Charles Dickens wrote his own bottlegram tale, 'A Message from the Sea' (1860), they were being laid between continents along undersea cables. Dickens's story, which he co-wrote with Wilkie Collins for a Christmas edition of *All the Year Round*, concerned a glass-stoppered medicine bottle discovered on an island reef by a Massachusetts sea captain, Silas Jorgan, who had been blown off course while rounding Cape Horn in a gale. The bottle contained a letter addressed to the younger brother of its shipwrecked author, whom Jorgan traces to their family home in Devon. In a plot twist supplied by Collins, the marooned

sailor is reunited with the message that he had consigned some
years before 'to the mercy of the deep':

> Thousands and thousands of miles away, I had trusted that
> Message to the waters – and here it was now, in my brother's
> hands! A chilly fear came over me at the seeing it again.
> Scrap of paper as it was, it looked to my eyes like the ghost
> of my own past self, gone home before me invisibly over the
> great wastes of the sea.[5]

As the literary historian Anthea Trodd observes, the story's
theme is 'the healthful circulation of information and sympathy',
with Captain Jorgan crossing the world to deliver a message
picked up in the Southern Ocean, as though he himself were
part of the great circulating system, or rather (given the speed
of delivery) one of the undersea telegraph cables that would
soon traverse it.[6] The idea of messages passing invisibly along
the seabed, rather than drifting on the surface, was evidently an
unsettling one, and Jorgan's actions emphasized the otherwise
ordinary human impulse that underlay the unfamiliar transoceanic technology.

By the end of the century, messages in bottles had become a
routine research tool, with thousands of annotated containers
hurled into the world's oceans for collection wherever they made
landfall. Between 1885 and 1887 Prince Albert I of Monaco, a keen
amateur oceanographer, released 1,675 bottles from his schooner
Hirondelle, while u.s. naval engineer George W. Melville designed
the 'drift cask', a strong wooden cask in the shape of a rugby ball,
fifty of which he released from Point Barrow, Alaska, between
1899 and 1901: one reached Siberia in 1902, another reached Iceland three years later, and a third was found on the coast of
Norway in 1908. They were the first known human-made objects
to traverse the Northwest Passage.[7] Unlike natural driftwood,
which can float for years over thousands of kilometres of sea,
ending up on distant shorelines far from home, labelled objects
are easily identifiable, even after many decades. In January 2018
a green glass bottle was picked up on a remote beach in Western

Australia; the still-legible message inside revealed that it had been thrown from a German barque, *Paula*, in June 1886, as part of a long-term experiment into ocean currents conducted by the German Naval Observatory. Thousands of drift bottles were thrown overboard during the seventy-year German survey, but only a few hundred were ever recovered. Given the prevailing currents and drift patterns of the southern Indian Ocean, where the bottle had been jettisoned, it is likely to have spent less than a year in the water before being buried by shifting sand on the Australian beach where it was discovered more than 130 years later: the world's oldest message in a bottle yet found.

Today, there is a network of nearly 4,000 subsurface 'Argo floats', whose buoyancy can be adjusted for travel at any depth in the ocean. The floats collect data on the temperature, salinity, oxygen and phytoplankton levels of the oceans, automatically rising to the surface every five to ten days to transmit their findings via satellite, before returning beneath the waves to continue their journey. What the wanderings of all these objects has shown is that, given time, all the world's oceanic waters are connected by currents, so even the water that is currently located 8 or 10 km (5 or 6 mi.) deep within a remote Pacific trench will, over the course of many hundreds of years, eventually find its way to a beach in Iceland, or India, or Antarctica, with every incoming or outgoing tide playing a tiny part in the great circulation of water and energy that distributes heat, nutrition and life to every corner of the planet.

The rise of oceanography

Oceanography is a surprisingly young science, with knowledge of the seas retaining a practical rather than a theoretical dimension for much of human history. Even the ancient Greeks, with their rich maritime culture, devoted little time to theorizing about the sea, with the exception of the perennial question of its saltiness, which had puzzled natural philosophers for centuries. A variety of competing answers had been proposed over the years, from Democritus' suggestion that the oceans have shrunk and thereby become more concentrated, to Empedocles' bodily analogy of the sea as the salty, extruded sweat of the earth. Aristotle developed Empedocles' idea, though with urine rather than sweat as the ruling analogy:

> Actually the saltness seems to be due to the same cause as in the case of the residual liquid that gathers in the bladder. That, too, becomes bitter and salt though the liquid we drink and that contained in our food is sweet. If then the bitterness is due in these cases (as with the water strained through lye) to the presence of a certain sort of stuff that is carried along by the urine (as indeed we actually find a salt deposit settling in chamber-pots) and is secreted from the flesh in sweat (as if the departing moisture were washing the stuff out of the body), then no doubt the admixture of something earthy with the water is what makes the sea salt.[8]

This was not a bad explanation, given that the saltness of the sea is due to rain washing mineral ions from the land into the water, but Aristotle was less accurate when it came to marine biology, arguing that there were only 180 species of sea creature (bizarrely, the Roman naturalist Pliny the Elder later amended this figure downwards, maintaining that there were in fact only 176 species to be found in all the world's oceans). Greek and Roman navigators had made practical use of sounding leads for measuring water depth, but only one attempt at a deep-sea fathoming was ever recorded, a 1,000-fathom (1,800 m) sounding of the Tyrrhenian Sea that was reported in the first century BC by the Turkish-born geographer Strabo.

Later centuries saw the development of more systematic forms of research, with European navigators beginning to compile and share reliable records of ocean currents and gyres. The 1580s saw the development of the modern nautical chart, with Dutch cartographer Lucas Janszoon Waghenaer's pioneering *Spieghel der zeevaerdt* ('Mariner's mirror') (1584) showing depth soundings and sailing directions across the waters of northwest Europe, as well as permanent hazards to shipping such as shoals and rocks.

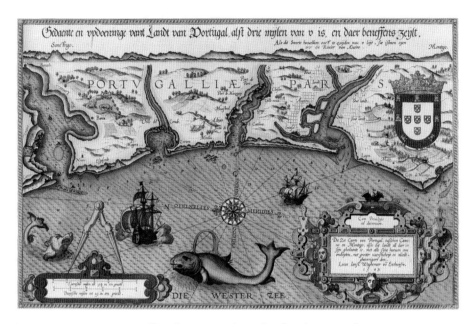

An early maritime map of Portugal, from Dutch cartographer Lucas Janszoon Waghenaer's *Spieghel der zeevaerdt* (Mariner's Mirror), 1584, the first modern nautical chart.

But the practical needs of explorers and navigators were not always aligned with more theoretical forms of scientific inquiry, and the tension between the two approaches did much to shape the early history of ocean science. In 1664, for example, the polymath and founding Fellow of the Royal Society of London Sir William Petty designed a double-hulled sloop (a variety of catamaran), which he named HMS *Experiment* in honour of the link between science and the sea. He had been interested in 'the philosophy of shipping' for some time, writing in 1661 that 'there was no greater, no more stately, no more useful, nor more intricate engine in the World than a Ship', and had earlier built a working prototype, the *Invention*, in which he completed a successful sea crossing between Dublin and Holyhead.[9] Samuel Pepys, the Navy Board's Clerk of the Acts, was a prominent supporter of Petty's double-hulled design, although established shipbuilders were wary, with one describing it as 'the most dangerous thing in the world' (a comment Pepys attributed to self-interest). The *Experiment*'s launch at Greenwich was a high-profile affair, with King Charles II and his brother the Duke of York in attendance, alongside the ubiquitous diarists,

Pepys and his friend John Evelyn. But Pepys's verdict on the vessel, that 'it swims and looks finely, and I believe will do well', proved premature, for six months later the *Experiment* capsized in a storm in the Bay of Biscay, and the Royal Society decided that maritime affairs, as a matter of state concern, were 'not proper to be managed by the Society', and they would no longer support shipbuilding projects of any kind.[10]

The episode illustrates some of the factors that inhibited the development of oceanography, not least the idea that maritime subjects, indeed the seas themselves, were the preserve of governments rather than professional institutions. Military and commercial interests dominated maritime affairs, but as early advocates of organized science such as the Royal Society argued, scientifically driven advances in navigation and mapping should not be overlooked. Robert Boyle's treatise *Observations and Experiments about the Saltness of the Sea* (1674), for example, explored means by which mariners could access fresh water during long sea voyages, while Robert Hooke delivered a series of lectures on methods in deep-sea research; though as the historian Susan Schlee observes, these inquiries tended to be concerned with one aspect of the sea at a time, 'and usually with such a one as could be poured into a jar and studied in the laboratory without further reference to the rest of the ocean'.[11]

Instrumental developments during the eighteenth century, however, saw European sea powers gradually begin to fund hydrographic surveys as a means of gaining commercial and imperial advantage. Alexander Dalrymple, hydrographer to the East India Company from the 1770s, was appointed hydrographer to the British Admiralty in 1795, but it was not until 1808 that a team of naval surveyors was actually recruited to make charts and take soundings of strategically important areas of sea. In the United States, too, hydrographic sounding only began in the 1830s, two decades after the founding of the u.s. Coast Survey. These early hydrographic efforts largely consisted of depth measurements taken by sounding pole and lead line paid out by hand, with positions determined by three-point sextant fixes on mapped reference points. While the initial

Dredging and
sounding equipment
on board the
converted survey ship
HMS *Challenger*, 1872.
Illustration from
C. Wyville Thomson's
*The Voyage of the
Challenger* (1878).

depth soundings were more or less accurate, they were limited in number, and so overall coverage was sparse. But these early surveys paved the way for more ambitious undertakings, and the *Challenger* expedition of 1872–6 became the first example of state-level 'big science', with a price tag to match (around £200,000, worth more than £20 million today). The Royal Society of London convinced the Royal Navy to lend them a three-masted corvette, HMS *Challenger*, and then persuaded the government to fund its modification for scientific use, rigging

her with a well-equipped dredging platform, along with dedicated research laboratories below decks, filled with specimen
jars, microscopes, thermometers, barometers, sampling bottles,
sounding leads and more than 300 km (190 mi.) of Italian-made
rope on which to send the equipment all the way down to the
ocean floor.

The expedition had been spurred, in part, by the spread of
the so-called azoic (or 'lifeless') theory, a term coined by the
British zoologist Edward Forbes, who had conducted small-
scale dredging experiments around the British Isles in the 1830s,
and later in the Mediterranean and Aegean seas in his role as
naturalist on board the surveying ship HMS *Beacon*. Forbes
dredged to depths of more than 400 m (1,300 ft) from which he
derived a hypothesis of eight bands or depth zones in the sea,
each having a unique population of marine life that diminished
with depth, so that life ceased, he argued, below 300 fathoms
(550 m/1,800 ft), 'pointing to a zero in the distribution of animal
life as yet unvisited'.[12]

It would not be long before Forbes's theory was proved
wrong. In 1860 a damaged submarine cable, which had been laid
between the Algerian coast and the island of Sardinia three
years before, was fished out of the Mediterranean for repair.
When it was brought to the surface, it was discovered to be
swarming with unfamiliar corals and clams dredged up from the
sea floor along with the cable, from a depth of 1,000 fathoms.
Subsequent dredging expeditions conducted from HMS *Lightning* and HMS *Porcupine* in the late 1860s discovered numerous
species such as clams, scallops and corals thriving at depths well
below Forbes's theoretical limit, and these results contributed to
the growing conviction that the open seas were not only an
unexplored terrain but an untapped resource.

The *Challenger* expedition marked the beginning of modern
oceanography, with historians even able to pinpoint 'the birth of
the science of the sea' to a specific date, 3 January 1873, when the
Challenger's onboard scientists made the first of many hundreds
of soundings and dredge hauls, just west of Lisbon.[13] The expedition's brief was to collect data and samples covering a wide array

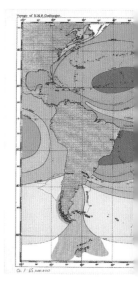

Colour-coded chart
of the *Challenger*
expedition, December
1872 to May 1876,
showing the dredging
depths achieved
throughout the oceans
of the world.

of ocean features, including water temperatures, saltwater chemistry and sea floor geology, as well as identifying new forms of marine life. Three and a half years and more than 130,000 km (70,000 nautical mi.) later, the vessel returned to the Solent, having conducted deep-sea research in the North and South Atlantic and Pacific oceans, and having travelled north of the limits of drift ice in the polar seas and south of the Antarctic Circle. Hundreds of notebooks had been filled with data, collecting bottles were crawling with amazing animals and plants, and numerous boxes were filled with rocks and mud hauled up from the deep-sea floor. 'The mud! ye gods, imagine a cart full of whitish mud, filled with minutest shells, poured all wet and sticky and slimy on to some clean planks,' as Sublieutenant George Campbell recalled of his time on the ship; 'in this the naturalists paddle and wade about, putting spadefuls into successively finer and finer sieves, till nothing remains but the minute shells, &c.'[14] The antics of 'the Scientifics', as the onboard naturalists were known among the vessel's regular crew, proved a source of bemusement as well as growing resentment, with the scientific work renamed 'drudging' by bored crew members who were anxious to be home after many months away at sea.[15]

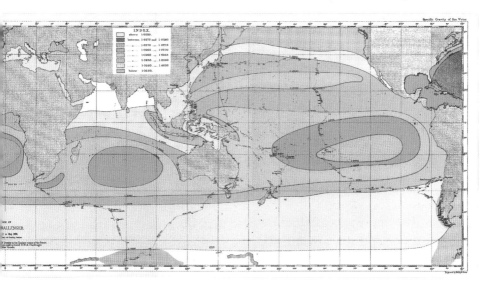

Ten years later, when the expedition's fifty-volume report was published, it featured never-before-seen details of the currents, temperatures, depths and constituents of the world's oceans, along with photographs, drawings and descriptions of the life forms of the abyssal waters. In all, more than 4,700 new species of marine life were identified, including anglerfish, Antarctic starfish, sea pigs (potato-like creatures that scuttle around the sea floor), and faceless cusk-eels, alongside hundreds of varieties of cephalopods and foraminifera. The naturalists had not expected to find such living bounty, and neither had they imagined how deep the ocean could be. At the lowest point recorded, near Guam in the western Pacific, the sea floor lay more than 8 km (5 mi.) below the surface, and the first time the crew sent down a sounding line, they ran out of rope; when they sent down a thermometer, the glass cracked from the enormous pressure. The scientists named the spot 'Swire Deep' after one of

The steam-powered ocean survey ship *Surveyor* passing through pancake ice in the Bering Sea, 1980s.

the ship's naval officers, but it has since been renamed the Challenger Deep, an abyss within the Mariana Trench where the sea floor rests nearly 11,000 m (36,000 ft) below the waves, in a little-explored region known as the Hadal Zone. The expedition also revealed the first broad outline of the shape of the ocean basin, including the mountain chain that extends the length of the Atlantic Ocean, now known as the Mid-Atlantic Ridge, a tectonic boundary between the diverging Eurasian and North American plates, and the longest mountain range on Earth.

The *Challenger* voyage was not exclusively scientific – the hydrographic office needed reliable information about the sea floor in advance of the transoceanic cable-laying campaign – but it resulted in the creation of an international community of marine scientists, and established the new science of oceanography as a subject of major collaborative research, bringing together areas of biological, geological and physical enquiry that had previously been undertaken separately.

The word 'oceanography' itself reflected the science's international character, having briefly appeared in French in the late sixteenth century (as *océanographie*), before reappearing in 1878 in the *Grand Dictionnaire Universel*, where it was defined as '*la description de l'océan*'. Contemporary German naturalists were using the term *Thalassographie* to refer to the study of gulfs and seas (derived from the ancient Greek *Thalassa*: the sea) but in the early 1880s the German-born chemist William Dittmar proposed the term 'oceanography' for use in scientific English, and by the end of the nineteenth century it was in widespread international use.[16]

Like space exploration, a branch of science with which it is often compared, deep-sea oceanography is an expensive and risky endeavour. In fact more people have been sent into space than have ever journeyed into the dark zone, 2,000 m (6,560 ft) below the surface of the ocean, and there are still only a handful of unmanned submersibles that are capable of reaching the deep-sea floor, 6 or 7 km (3–4 mi.) down.

The danger of deep-sea exploration arises from the steady increase in hydrostatic pressure that is caused by the weight of water above. By 1,000 m (just over half a mile) down, the outside pressure will have risen to a hundred times the pressure experienced at the surface. By 5,000 m (just over 3 mi.) down, the pressure will have increased to 500 atmospheres – around half a tonne per square centimetre (3.5 tons per square inch) – inducing levels of stress that few human-made objects can withstand. A polystyrene coffee cup strapped to the outside of a submersible would be slowly crushed to the size of a doll's-house thimble as the craft makes it descent.

Questions relating to ocean depth were difficult for scientists to answer in the years before the advent of sonar imaging. Soundings taken at the same coordinates could vary from ship to ship, and it was difficult to judge when the sea floor had been reached, since undercurrents tended to pull lines from their reels long after the sinkers had struck bottom. Consequently, a great deal of time at sea was spent experimenting with a range of sounding equipment in pursuit of a reliable 'law of descent' that would let the user know exactly when the sinker, a 14-kilogram (32 lb) shot attached to a specific diameter of sounding line, had landed. Every piece of kit, including the lines, would be tested and retested over the years, with hemp ropes and steel cables giving way in the end to cheap commercial packing twine, the only kind of line that did not break when hauled from a deep-sea sounding. As marine historian Helen Rozwadowski observed, the story of the oceans during the mid-nineteenth century was 'not merely the pencilling in of a previously blank chart', but a more three-dimensional apprehension of ocean depths, an 'expanding imagining of what the "deep sea" might be'.[17]

The early twentieth century saw the development of single-beam echo sounders that used sound to measure the distance of the sea floor directly below a vessel. By running a series of lines at a specified spacing, echo sounders and fathometers increased the speed of the survey process by allowing more data points to be collected. However, this method still left gaps in quantitative depth information between survey lines. Later innovations such as side scan sonar technology offered a means of obtaining the sonic equivalent of an aerial photograph, improved the ability to identify submerged wrecks and obstructions, while more recent multibeam systems have made it possible to obtain quantitative depth information for the whole sea floor beneath any given survey area: technology that is a world away from the hand-hauled ropes and buckets used on the *Challenger* expedition and its forerunners a century and a half before.

Waves and tides

Sea waves are among the world's most misunderstood phenomena. When an incoming wave breaks on the shoreline, it appears as though the water has come to the end of a long journey, but in fact the water itself has hardly moved. Most surface sea waves transmit energy, not water, and the turbulence at the surf zone is the result of that moving energy encountering a solid obstruction – usually the shelving sea floor – against which it noisily dissipates. It is at that point that the wave transforms, from an energy-transporting wave of oscillation to a water-moving wave of translation, commonly known as 'swash'. So for most of its life a wave is not a thing so much as an event, a small part of a large-scale transfer of energy from one part of the sea to another.

Waves are most commonly generated by the friction between wind and the surface of the water. As wind blows across the sea, the disturbance and perturbation cause a small wave crest to form, and the resulting up and down motion begins to transmit kinetic energy through the water in the form of a series of waves. As the waves grow, the energy (but not the water) passes from crest to crest, as is apparent when a piece of flotsam, say a tin can, can be seen bobbing up and down on the spot as waves pass beneath it.

Waves are classified according to their wave period or wavelength (the distance between two crests), from the smallest capillary waves to the greatest waves of all, the tides, and those who study them are known as kumatologists, from the Greek *kumas* ('wave'), a term coined by the wave-obsessed English geographer Vaughan Cornish in 1899. Capillary waves are the

small rippled disturbances that first appear on the surface of wind-blown water, which have been known to mariners for centuries as 'cats' paws'. William Henry Smyth, in his *Sailor's Word-Book* (1867), defined the cat's paw as 'a light air perceived at a distance in a calm, by the impressions made on the surface of the sea, which it sweeps very gently', and noted the widespread superstition of rubbing a ship's backstay to invoke the lucky cat's paw, 'the general forerunner of the steadier breeze'.[18] Samuel Taylor Coleridge, en route to Malta in April 1804, wrote a finely observed description of the various wave types seen from deck, starting with the hair's-breadth ripples of capillary waves:

Powerful wind-driven waves breaking in the surf zone off the coast of Pacifica, California.

I particularly watched the beautiful Surface of the Sea in this gentle Breeze! – every form so transitory, so for the instant, & yet for that instant so substantial in all its sharp lines, steep surfaces, & hair-deep indentures, just as if it were cut glass, glass cut into ten thousand varieties / & then the network of the wavelets, & the rude circle hole network of the Foam.[19]

As is evident from both Smyth's and Coleridge's accounts, the characteristically rippled structure of capillary waves is due to light breezes (blowing at speeds of about 3–4 m (10–13 ft) per second) that generate wavelengths typically less than 1.5 cm. Such gentle wind is not what sailors need to fill their sails, however, and neither is it enough to cause travelling waves to form. The threshold wavelength at which surface waves begin to travel is above 1.7 cm: anything shorter than that will be suppressed by gravity. But if – as Smyth's sailors would have hoped – the wind strengthens to blow consistently over a substantial fetch of water, the second class of sea wave, gravity waves, will then begin to form.

Gravity waves occur when wavelength grows to around 1.5 m (5 ft), and gravity joins forces with wind as the main dispersing agent. A slight convexity in the wave shape is needed to give the wind something to work on, and as soon as a wavelet develops a leeward face and a windward back, the wave (as it now is) will begin to climb: the water's line of least resistance is to go upwards as the energy in the wind is transferred to the sea. When longer gravity waves propagate over deep water, they move rapidly away from the generating wind, at which point they are known as swells, with a typical wavelength greater than 260 m (855 ft), up to a maximum of 900 m (2,950 ft). Swells lose so little energy as they cross open water that it's possible for one generated in the Antarctic Ocean to travel all the way to Alaska at full strength, taking several days to do so. Swells tend to flatten out as they move away from their source, and ripples and waves can form on top of underlying swells, causing a complex surface pattern which takes an experienced navigator to discern.

But even the longest-lasting swell will eventually approach a shoreline, at which point contact with the shoaling sea floor will cause it to break and dissipate its energy remarkably quickly.

Estimating the height of waves at sea has always been a challenge, given the lack of visual definition between moving crests and troughs. In the mid-1880s the Hon. Ralph Abercromby, Fellow of the Royal Meteorological Society, spent part of his inheritance travelling the world making meteorological and oceanographic observations. Upon his return he published a remarkable memoir, entitled *Seas and Skies in Many Latitudes; or, Wanderings in Search of Weather* (1888), which was a masterclass in sustained maritime observation. One of his particular interests was the shape and dimension of sea waves, and he noted many discrepancies in wave-height estimates among crew members on board ship. He concluded that it was impossible for one person to accurately measure the height of a given wave train, a process that would involve taking a series of lengths while on deck, then running down to a porthole below to take a series of corresponding heights. It required a minimum of three observers, he thought, 'one for the barometer, one for the chronographs, and a third to estimate the height of the deck from the water and record the readings of the various instruments', an idea that he went on to explore at length in a later publication.[20]

Wave heights out at sea remain challenging to measure, given that the sea's surface is composed of waves of varying heights and periods moving in differing directions, the simple matter of gauging the amplitude from crest to still-water line being impossible when the water is far from still. Such is the complexity of a wind-driven sea that modern wave-recording equipment (pressure sensors and buoys) do not give readings of individual wave heights, but instead generate statistical or probabilistic readings based on samples. The pressure sensors are mounted at fixed positions underwater, measuring the height of the wave column that passes above. As wave crests pass by, the height of the column increases; when troughs approach, the height of the column falls. By deducting the depth of the sensor

from the water-column heights, an averaged record of sea surface elevations is generated, though this is not the same as
measuring the height of an individual wave.

The longest waves of all are the tides, the vertical movements of seawater dragged by the gravitational pull of the moon
and sun. High tides and low tides occur on opposite sides of our
planet at the same time, and, in the words of navigator Tristan
Gooley, 'can be thought of as a pair of very long waves that ride
around the Earth', straddling half the globe between crests.[21]
Most coastal areas experience two high tides and two low tides
every lunar day – the time taken for a specific spot to rotate from
and return to an exact point under the Moon: a lunar day is fifty
minutes longer than a solar day because the Moon revolves
around Earth in the same direction that Earth rotates around
its axis, so it takes Earth a little more time to 'catch up' with the
Moon. The section of ocean nearest the Moon is pulled by gravity into an upward bulge, while the Earth's rotation sends that
bulge moving round the planet, giving us two high and two low
tides every 24 hours and 50 minutes.

Tides are global forces, but they also feature local and
regional improvisations – eddies, whirlpools, vortices, mael-
stroms – their cosmic regularity perturbed by wind and weather
conditions, as well as by local topography. Gravitational lift
alone can only raise a tide of some 45 cm (18 in.): anything
higher than that is caused by the shape of the coastal land and
its influence on the high-tide bulge, which can be funnelled into
heights as great as 15 m (50 ft), as seen in the steeply shoaling
Severn Estuary. When a tide makes contact with the coast, its
water rushes up inlets and estuaries, reflects off sea walls, swirls
around rocks and islands. It can even bounce back and cancel
itself out, as happens in parts of the Gulf of Mexico, where there
is only one high tide per day, while the opposite effect can lead
to so-called 'double tides', where the topography and resulting
flow leads to a double peak at high tide – an extra-long period
of high water – which is the main reason why Southampton,
shielded by the Isle of Wight on England's south coast, was
developed as a major naval and commercial port.[22]

A Coast and Geodetic Survey officer checking a tide gauge installed off the coast of West Greenland, 1926.

But beneath their capriciousness, tides express an under-lying order that is almost clock-like in its regularity, with a twelve-and-a-half-hour tidal cycle governing most of our plan-et's coastal waters: tides typically run for around six hours in one direction, before reversing and running in the opposite direction for the following six hours. The writer Hugh Aldersey-Williams has described this as 'one of the earth's fundamental

units of time', a temporal precondition that has made itself at home in our language, with its ancient interplays of time and tide.[23] The Anglo-Saxon word *tíd* (from the proto-Germanic *tīdiz*) denoted both the sea's rhythmic motions and times of religious significance, with the latter sense retained in calendrical terms such as Shrovetide and Whitsuntide. In later centuries these meanings began to separate, with tides more often spoken of as *flód* (flood) and *ebba* (ebb), alongside an emerging specialist terminology including *apflód* (low tide), *héahflód* (high tide), *népflód* (neap tide) and *fylleflód* (spring tide: though 'spring', in this case, refers to an upwelling of water rather than the season, to which the term, surprisingly, has no connection; the medieval adoption of the word has misled people ever since into expecting higher tides in spring). The Old English *heahtid* (high tide), meanwhile, referred to festivals

A rogue wave looms before a merchant ship in the Bay of Biscay, *c.* 1940. Rogue waves are believed to form from a series of waves joined together through a combination of strong winds and fast currents.

(or 'high days'), while the adjective 'tidy' also derived from *tíd*, its original sense of 'seasonable' or 'in its place' still discernible in the modern meaning. The figurative uses of the tidal terms *flood* (as in a flood of tears) and *ebb* (as in something dwindling away) had been established in English by the fifteenth century, while other uses have changed their meaning over time: to tide someone over, for example, today implies financial support, whereas the original, nautical, meaning referred to using a period of high water to 'tide' a vessel and its navigator over a sandbar or other shallow impediment that would be impassable at other states of the tide.[24]

Understanding tides was vital for seafarers, and the first predictive tide table appeared in eleventh-century China in response to the powerful and deadly tidal bore that surges up the Qiantang River. Known locally as the Black Dragon, the world's largest tidal wave attracts thousands of spectators, but the tide tables outlining wave times and heights can be notoriously inaccurate. In August 2013, the bore was roughly twice its predicted height, sweeping hundreds of spectators into the water, while the following year the appearance of a 'supermoon' generated another unexpectedly powerful 9-metre (30 ft) wave that injured dozens more.

Further out at sea, tide-related perturbations such as whirlpools, vortexes and tidal races are not always marked on charts. 'They are places, but they are also theatrical events,' as Hugh Aldersey-Williams observed in the course of an account of his journey into the maelstrom, a large whirlpool off Norway's Lofoten Islands, known locally as the Moskstraumen, after the nearby Moskenesøya Island. Rather than the wild vortex of literary legend, as invoked by Edgar Allan Poe in his celebrated story *A Descent into the Maelström* (1841), as well as by Jules Verne in the tragic finale of *Twenty Thousand Leagues Under the Seas*, the phenomenon is a system of tidal eddies and whirlpools that exude, in Aldersey-Williams's words, 'a quiet venomous authority in contrast to the petulant rage of great waves'.[25]

There have been a surprising number of variant spellings of the Norwegian word *maelstrom*, from *malen* (to grind) and

stroom (stream), including Poe's 'Maelström', Jules Verne's 'Maël-strom' and A. S. Byatt's 'Maelstrøm' (in which the subject of her comic novel *The Biographer's Tale* (2000) meets a Nemo-like death), the chaotic typography paying a kind of linguistic tribute to the imagined violence of the phenomenon.[26] Poe's story, which introduced the word 'maelstrom' into the English language, featured a bravura, if highly exaggerated description:

The maelstrom off the coast of Norway, as illustrated in Olaus Magnus's *Carta Marina*, 1539.

> Even while I gazed, this current acquired a monstrous velocity. Each moment added to its speed – to its headlong impetuosity. In five minutes the whole sea, as far as Vurrgh, was lashed into ungovernable fury; but it was between Moskoe and the coast that the main uproar held its sway. Here the vast bed of the waters, seamed and scarred into a thousand conflicting channels, burst suddenly into phrensied convulsion – heaving, boiling, hissing – gyrating in gigantic and innumerable vortices, and all whirling and plunging on to the eastward with a rapidity which water never elsewhere assumes except in precipitous descents.

'A Descent into the Maelstrom': illustration by Harry Clarke from a 1919 compilation of Edgar Allan Poe's writings, *Tales of Mystery and Imagination*, showing the vortex down which the short story's narrator's boat is dragged to its doom.

Not long after, the violence of the waters diminished, and the whirlpools disappeared, to be replaced by 'prodigious streaks of foam' that spread over a great distance before beginning a curious gyratory motion:

> Suddenly – very suddenly – this assumed a distinct and definite existence, in a circle of more than a mile in diameter. The edge of the whirl was represented by a broad belt of gleaming spray; but no particle of this slipped into the mouth of the terrific funnel, whose interior, as far as the

Utagawa Hiroshige,
'Naruto Whirlpool,
Awa Province', from
*Views of Famous
Places in the Sixty-odd
Provinces* (*c.* 1853).

eye could fathom it, was a smooth, shining, and jet-black wall of water, inclined to the horizon at an angle of some forty-five degrees, speeding dizzily round and round with a swaying and sweltering motion, and sending forth to the winds an appalling voice, half shriek, half roar, such as not even the mighty cataract of Niagara ever lifts up in its agony to Heaven.[27]

Though whirlpools can be noisy and violent – they are caused when fast-moving tidal currents are sent into a spin by local topography, or by the meeting of a pair of opposing currents – the original, Norwegian, maelstrom is fairly pacific. It features a sequence of choppy waves and declivities fuelled by strong diurnal tides rather than the hole in the sea – 'the navel of the ocean' – implied by the popular legends.[28]

But Earth will not experience such phenomena for ever. A billion or so years ago, the Moon was much nearer than it is today, with the correspondingly stronger gravitational attraction raising vast tides in Earth's young oceans that had condensed from atmospheric water vapour. The energy dissipated by those tides as friction helped slow Earth's rotational and orbital speeds, locking our Moon into its present alignment, so it always shows the same face to Earth. Eventually our slowing planet will also become tidally locked, and so always show the same face to the Moon. The tides will then cease, depriving any surviving intelligent life of one of Earth's great natural dramas, the spectacle of the oceans' rise and fall under the pull of the Moon's cosmic leash.

The Gulf Stream

Among the many irritations suffered by the British during their administration of the American colonies was the fact that ships dispatched from Britain took two weeks longer to cross the Atlantic than ships dispatched from America. At a loss to account for this disparity, which could not be attributed to prevailing winds, the Admiralty eventually approached Benjamin

Franklin, scientist and Postmaster General to the colonies, whose subsequent enquiries among his seafaring contacts revealed what Nantucket whalers had known for generations: the existence of a swift Atlantic current that plies its way along the eastern seaboard before heading across the open sea to Europe. Franklin, whose next two crossings were spent testing the waters, confirmed that a fast-flowing ocean river, known as the Gulf Stream, did indeed run across the North Atlantic, and that 'a stranger may know when he is in it, by the warmth of the water, which is much greater than that of the water on each side of it'.[29] It was during an Atlantic crossing in April–May 1775 that he devoted most time to the subject, lowering his thermometer into the ocean several times a day, from seven in the morning to eleven at night, while the ship kept along the eastern edge of the stream and then cut across it into colder water. He noticed that the water in the stream had a different colour from the surrounding seawater, partly due to the increased quantity of gulf weed to be found in it, and that 'it does not sparkle in the night'. But it was his insight that the Gulf Stream 'might be studied like a river' that transformed contemporary understanding of ocean currents, while pointing out the means by which vessels, armed only with a thermometer, could shorten their crossings in both directions by keeping variously in and out of this flowing Atlantic river.[30]

Research into the movements of the Gulf Stream was continued by Alexander Dallas Bache, a superintendent of the United States Coast Survey, who also happened to be Franklin's great-grandson. In 1845 Bache issued a set of instructions for Gulf Stream studies that did much to establish the framework of the modern oceanographic survey:

> *First.* What are the limits of the Gulf Stream on this part of the coast of the United States, at the surface and below the surface?
>
> *Second.* Are they constant or variable, do they change with the season, with the prevalent and different winds; what is the effect of greater or less quantities of ice in the vicinity?

Benjamin Franklin's chart of the Gulf Stream, *c.* 1782. Franklin spent several Atlantic crossings tracking the course of the Gulf Stream through regular temperature readings.

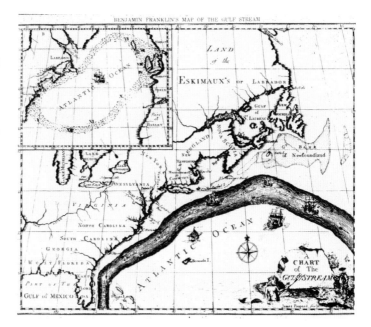

Third. How may they best be recognized, by the temperature at the surface or below the surface, by soundings, by the character of the bottom, by peculiar forms of vegetable or animal life, by meteorology, by the saltness of the water?

Fourth. What are the directions and velocities of the currents in this Stream and adjacent to it at the surface, below the surface, and to what variations are they subject?[31]

By bringing together a range of physical, chemical, geological, biological and meteorological questions, Bache extended oceanography towards an understanding of the ocean as a complex system of physical and biological interrelationships. Bache was proud to have continued the oceanographic work begun by his great-grandfather, although it would cost the life of his brother, George, a naval officer in command of the survey brig *Washington*, who was swept overboard along with ten of his crew during a storm off Cape Hatteras while engaged in Gulf Stream research in September 1846.

In the centuries since, the course and pace of the Gulf Stream have been tracked and measured, as have the changes that it undergoes during its long Atlantic crossing: as the current heads north, its cargo of warm tropical water cools and evaporates, leaving saltier, heavier water behind. By the time it reaches the Norwegian Sea the Gulf Stream's payload is cold and dense enough to start sinking towards the ocean floor, turning to flow south towards the equator in a cold counter-current that will eventually return it to the point where its journey began. The effect of this great circulation is to transfer significant amounts of heat across the North Atlantic – a thermal budget almost one hundred times greater than world energy demand – preserving

The u.s. Coast Survey brig *Washington* encountered a severe storm in September 1846 while engaged in Gulf Stream studies. Lt George M. Bache and ten of his crew were swept overboard to their deaths.

the inhabitants of western Europe from the harsh winters which besiege other places at similar latitudes, such as Canada or southern Alaska.

The process of thermohaline warming was first described in the 1850s by the American hydrographer Matthew Fontaine Maury, who characterized the oceans as a vast and efficient boiler house (the term derives from *thermo* (heat) and *haline* (salt), the two factors that influence the density of seawater). As Maury outlined in his landmark study *The Physical Geography of the Sea* (1855), 'the furnace is the torrid zone, the Mexican Gulf and Caribbean Sea are the caldrons; the Gulf Stream is the conducting pipe' distributing tropical warmth towards Britain and western Europe. 'It is the influence of this stream upon climate that makes Erin the "Emerald Isle of the Sea", and that clothes the shores of Albion in evergreen robes, while in the same latitude, on this side, the coasts of Labrador are fast bound in fetters of ice,' he wrote:

> There is a river in the ocean; in the severest droughts it never fails, and in the mightiest floods it never overflows. Its banks and bottoms are of cold water, while its current is of warm. The Gulf of Mexico is its fountain, and its mouth is in the Arctic Seas. It is the Gulf Stream. There is in the world no other such majestic flow of waters.[32]

Maury's rhapsodic description captures the vastness of the Gulf Stream phenomenon, while hinting at its widespread influence on Earth's weather and climate. Such thermohaline currents are the engines of Earth's weather, due to seawater's efficiency at storing and transporting solar heat. Water covers 70 per cent of the planet's total area, with around 97 per cent of that water contained in the oceans; most of the remainder is locked up in ice sheets and glaciers, and less than 0.001 per cent is ever present in the atmosphere: enough for only ten days' rain. Thermohaline circulation transports a massive current of water around the globe, from northern oceans to southern oceans and back again. These currents – driven by differences in water density – slowly turn seawater over in the ocean, from top to bottom, like a vast conveyor belt, sending warm surface water downwards and forcing cold, dense, nutrient-rich water upwards. Cold seawater in the polar regions forms sea ice, as a consequence of which the surrounding water becomes saltier, since when sea ice forms, the salt is left behind. As the polar water grows saltier, its density increases, and it starts to sink. Surface water is then pulled in to replace the sinking water, which in turn grows cold and salty enough to sink, too. This movement is what initiates the deep-ocean currents that drive the global conveyer belt, with water always seeking an equilibrium; when cold, dense water sinks, sending warmer water welling up from below to balance out the loss at the surface.

This results in an intermixing of the solar energy collected by the top layer of ocean and the nutrient-filled sediment of decayed plant and animal matter found at the bottom. Without the great conveyor belt stirring them up, along with the wind-driven upwelling that occurs near coastal regions, most of these nutrients would remain sequestered at the sea floor, leaving the oceans unable to support the dazzling array of life forms (perhaps as many as 2 million species, the majority of which have yet to be classified or named) whose mysterious, complex and inter-related ecologies will be the subject of the following chapter.

3 Sea Life

The sea would eat the sea if it had jaws.
Jeremy Reed, 'Dogfish', in *By the Fisheries* (1984)

The largest migration of life on Earth departs every night from
the 'twilight zone', the kilometre-deep middle layer of open ocean
in which the majority of living creatures can be found. As darkness falls across the seas, millions of tonnes of animals, ranging
from the smallest arrow worms to the largest cetaceans, swim
their way up to the 'sunlight zone' to feed in relative safety, brav-
ing shallower waters under cover of night to gorge themselves on
krill, zooplankton and microscopic phytoplankton (as well as on
one another) before plunging back into the twilit depths as dawn
begins to break. 'The setting of the sun is a difficult time for all
fish,' observes the narrator of Ernest Hemingway's *The Old Man
and the Sea* (1952), and throughout the hectic hours before sunrise,
the top 30 m (100 ft) of the world's great oceans teem with life
like overstocked aquaria, in which some 3.5 billion tonnes of
phytoplankton are consumed each night: nearly eight times the
weight of all the people on the planet.[1]

The process, known as vertical migration, is of relatively
recent discovery, and as yet scant details of its natural history
have been collected by marine zoologists, for whom some of the
creatures in the ocean's deeper realms remain just as mysterious
as they were when ocean science began in earnest in the
mid-nineteenth century. 'The great depths of the ocean are
entirely unknown to us,' as Jules Verne's narrator, Professor
Aronnax, declared in the opening pages of *Twenty Thousand
Leagues Under the Seas*: 'Soundings cannot reach them. What
passes in those remote depths – what beings live, or can live,

'The setting of the sun is a difficult time for all fish': moonlit sea, off Walberswick, Suffolk.

twelve or fifteen miles beneath the surface of the waters – what is the organisation of these animals, we can scarcely conjecture.'[2] And much the same could be said today, 150 years on from the submarine voyage of the *Nautilus*, with less than 5 per cent of the world's 1.3 billion cubic km (326 million cubic mi.) of ocean having so far been explored, and as many as 2 million unknown species living undiscovered in its depths ('depths' being key: the oceans cover around 70 per cent of the world's surface, but they provide more than 90 per cent of its living space). But our knowledge of the oceans is gradually increasing, and the advent of roving undersea technologies such as Remotely Operated Vehicles (ROVs) and Autonomous Underwater Vehicles (AUVs) promises a new and illuminating era of deep-sea exploration.

There are five distinct layers in the ocean: the sunlight zone, the twilight zone, the midnight zone, the abyss and the trenches (or, to give them their scientific names, the epipelagic, mesopelagic, bathypelagic, abyssopelagic and hadalpelagic zones), the last two being largely unexplored realms, where temperature drops and pressure increases to astounding levels: at only 300 m (900 ft) below the surface, at the top of the twilight zone, pressure is already thirty times that of Earth's atmosphere – enough to rupture organs in a human body. By 1,000 m (3,280 ft) down, at the top of the midnight zone, human lungs would already have collapsed. Most deep-sea creatures have adapted to withstand the enormous barometric pressures of the sub-photic zones, eliminating excess cavities, such as swim bladders, that would otherwise compress, and developing squishy, pressure-absorbing flesh like that of the elusive blobfish (*Psychrolutes marcidus*), which may have been voted the world's ugliest animal, but is nevertheless beautifully adapted to its deep-water environment, as is the mighty sperm whale (*Physeter macrocephalus*), which can dive to a depth of nearly 3 km (10,000 ft), a record for a mammal: at such depths their lungs will compress to flatness, while inside their heads are large chambers in which the pressure is equalized by the waxy, semi-liquid spermaceti oil – from which the sperm whale derives its name – cooling and solidifying to allow greater density during these creatures' lengthy descents into the deep.

These oceanic depths are pitch dark, so photosynthesis using energy from the Sun is no longer a viable source of nutrition. There is not enough light. But since many deep-sea species produce their own bioluminescent lights, the abyssal waters wink and flash with millions of spots of colour. Given the range of species that live in the deep ocean – far more than live on land – these chemically produced light shows constitute the most widespread form of communication on the planet, from the dangling fishing lure of the humpback anglerfish (*Melanocetus johnsonii*), which protrudes from its snout while it waits in ambush for its prey, to the bell-shaped atolla jellyfish (*Atolla wyvillei*), which flashes a diversionary series of bright blue lights when attacked, hence its nickname, 'the alarm jellyfish'.

Our seas are unimaginably rich underwater landscapes of currents and tides, shallows and depths, characterized by steep gradients of temperature, density and salinity, and populated by such an abundance of life that most of it remains unidentified. But though there is much that we don't yet know about the networks of life that are variously thriving and declining in our oceans – around 2,000 marine species new to science are

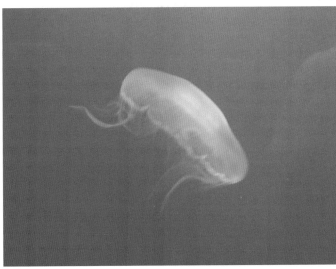

Moon jellyfish
(*Aurelia aurita*)
floating in the ethereal
blue of a public
aquarium.

discovered and named every year – this chapter will look at some of the more familiar forms of marine life, starting at the shore-lines, with their rock pools and shallows, before moving out into open water in search of the some of the lesser-known denizens of the deep.

Shore life

A memorable scene in *Father and Son* (1907), Edmund Gosse's unsparing account of his 1850s childhood, recalls the single-mindedness with which Gosse's widowed father, Philip, took to the then-nascent fashion for rock pooling. On their regular trips to the Devon coast, Gosse senior would 'wade breast-high into one of the huge pools, and examine the worm-eaten surface of the rock above and below the brim' in search of corals and anemones to add to his collection.[3] But thousands of others were soon doing the same, and by the end of the century, according to Gosse, the rock pools of Britain – 'the ring of living beauty drawn about our shores' – had been all but stripped of life:

> An army of 'collectors' has passed over them, and ravaged every corner of them. The fairy paradise has been violated . . . No one will see again on the shore of England what I saw in my early childhood, the submarine vision of dark rocks, speckled and starred with an infinite variety of colour, and streamed over by silken flags of royal crimson and purple.[4]

The tragedy for Philip Gosse was that he knew how much he had contributed to this despoilation, having published a series of popular books on marine biology, beginning with *A Naturalist's Rambles on the Devonshire Coast* (1853), in which he encouraged the collection of shoreline specimens for home study and display. The book described dozens of species in loving detail, along with the techniques needed to prise them from their habitats. 'The lovely Daisy Anemone', for example, might look simple to collect, but was surprisingly resistant to removal:

Sperm whale mother and calf off the coast of Mauritius. Sperm whales can dive to depths of nearly 3 km (10,000 ft), a record for any mammal.

'no sooner do the fingers touch one, than its beautifully circular disk begins to curl and pucker its margin, and to incurve it in the form of a cup . . . Nothing will do but the chisel, and this is by no means easy of appliance.'[5]

But it would be Gosse senior's next book, *The Aquarium: An Unveiling of the Wonders of the Deep Sea* (1854), that inaugurated the craze for indoor aquaria (the word was his own coinage), and led to the mass stripping out of rock pools and sea caves the length of the British coast. Gosse himself had been tasked with collecting anemones and other specimens of marine life for the first public aquarium, which opened at London Zoo in 1853, and

Rock pools, 'the ring of living beauty drawn about our shores', as Edmund Gosse described them in his coastal memoir, *Father and Son* (1907).

in *The Aquarium* he encouraged his readers to replicate this new exhibition in miniature:

> Any visitor to the sea-side, though there for ever so brief a stay, may enjoy, with the least possible trouble, the amenities of zoological study in a soup-plate, or even in a tumbler. It is easy to knock off with a hammer, or even to dislodge with a strong clasp-knife, a fragment of rock on which a minute sea-weed is growing, proportioning the surface of leaf to the volume of water, – and you have an Aquarium.[6]

This was followed by detailed instructions for the removal of all manner of plants and creatures, from periwinkles to crabs, most of which were fated to die in neglected or improperly cleaned tanks in the drawing rooms of suburban houses. Gosse even recommended hiring a boat for dredging trips out at sea, rhapsodizing over the abundant catches that might spill onto the deck: 'What a pleasure it is to examine a tolerably prolific dredge-haul! I am not going to enumerate all the things that we found; it would make a pretty long list', though we learn that

The ocean at home: aquariums became popular in the mid-19th century. The word was coined by Philip Henry Gosse in his 1854 book *The Aquarium: An Unveiling of the Wonders of the Deep Sea.*

the list included zoophytes, molluscs, crabs, prawns, shrimps, worms, sponges, seahorses and starfish of all kinds, from brittle stars to sand stars, the latter 'very difficult to keep alive', and many of which are now classed as vulnerable species, protected from collection in UK waters under the Wildlife and Country-side Act of 1981.[7]

For Gosse, a devout member of the Plymouth Brethren, collecting marine animals was a means of studying God's hidden handiwork, and he took great pleasure in uncovering the secret life of the seas as it revealed itself in his glass and birchwood tanks. He was particularly taken by the phenomenon of bio-logical symbiosis, in which the lives of two species become entwined to their mutual benefit. The sea anemone, for example, with its colour-changing tentacles and flower-like appearance, had long been considered a 'missing link' between plants and animals (it is now classified as an animal), but it was the sym-biosis between certain species of anemone and hermit crab that was the greater mystery: was one of these closely melded bodies male, the other female? And which was in charge – was it the crab, pulling the brightly coloured anemone along the ocean bed, or had the anemone released some kind of powerful mem-brane, trapping the scuttling crab inside the home it could never leave?

It was not until the 1930s that a plausible explanation arose for this close interaction, which is now regarded as 'the prime example of biological symbiosis'.[8] A sea anemone uses its cni-docytes (stinging cells) to keep predators such as octopuses away from the crab, while the usually stationary anemone benefits by gaining a roving habitat, courtesy of its mobile host. It can also ingest some of the crab's prey, since its mouth directly faces that of the crab – although of course, in times of food shortage, the crab can always turn on its trusting companion, and eat the poor anemone instead.

A similar fascination with marine symbiosis had been expressed by Benjamin Franklin, who, during a two-month Atlantic crossing during the autumn of 1726, observed the mutually beneficial relationship of sharks and pilot fish: 'A shark

Colour plate from Ernst Haeckel's *Kunstformen der Natur* (Art Forms in Nature), 1904, showing sea anemones of the genus *Actiniae*.

is never seen without a retinue of these', he wrote, 'who are his purveyors, discovering and distinguishing his prey for him; while he in return gratefully protects them from the ravenous hungry dolphin.'[9] For Franklin, as it would later be for Gosse, such symbiotic arrangements were clear evidence of a divinely ordered natural harmony in action.

According to his son, Gosse senior suffered 'great chagrin' over his encouragement of shoreline collecting, the consequences of which he had not foreseen, although in recent decades marine conservation has enacted a reversal of the journey from ocean to indoors that began in the Victorian era, with underwater aquaria being established out at sea in the form of artificial reefs.[10] Since the 1980s several thousand decommissioned New York subway carriages have been towed out on barges and dumped into Atlantic waters off eastern seaboard states from New Jersey to South Carolina, in order to supply three-dimensional habitats for vulnerable crustacea and fish, including sea bass, tuna, mackerel and flounder. One of the sites off the coast of Delaware, now known as Redbird Reef (after New York's iconic Redbird subway car), has seen a four hundred-fold increase in marine life in the years since more than seven hundred steel carriages were dumped onto the sandy ocean floor as luxury accommodation for fish. Mussels, in particular, benefit from the wide surface area provided by the subway cars, and millions of blue mussels (*Mytilus edulis*) now carpet these artificial reefs, inside and out. These mussels support a further thirty or forty marine species, including shrimps, crabs and worms, whose livelihood depends on the proximity of thriving bivalve populations.

The flash of fin

One of the best-known tag lines in cinema history appeared in the trailer for *Jaws 2* (dir. Jeannot Szwarc, 1978): 'Just when you thought it was safe to go back in the water.' The original *Jaws* (dir. Steven Spielberg, 1975) had been based on Peter Benchley's 1974 novel, written, according to Benchley, in response to a news report that big game fisherman Frank Mundus had landed a 4,550-pound (2,060 kg) great white shark off the coast of Long Island. Benchley struggled over the title for his novel, moving from 'Silence in the Water' to 'The Jaws of Leviathan', before realizing that the latter title, stripped back to the essential noun, spoke most directly to an atavistic human fear of sharks.

Sharks are an ancient order of fish, dating back more than 420 million years, though they continued to evolve until relatively recently; two of the nine species of 'walking' sharks are less than 2 million years old. They are found in all seas and in all sizes, ranging from the dwarf lanternshark (*Etmopterus perryi*), a deep-sea species measuring less than 20 cm (7 in.) in length, to the mighty whale shark (*Rhincodon typus*), the largest fish in the world, which can grow as long as 18 m (60 ft). Many shark species are apex predators in their marine environments, which partly explains the fear and respect in which they are held, and offers a clue to the etymology of their name, which derives from the sixteenth-century German word *Schorck* (sometimes *Schurke*), meaning 'villain', or 'scoundrel'. The related meaning of shark in English, 'a dishonest person who preys on others', was first recorded in 1599, although 'sharker' ('an artful swindler') is of earlier usage, and may have supplied the word's original sense, later applied to the fish due to its perceived predatory cunning.

The earliest recorded use of the word 'shark' in English dates from 1569, when a large Atlantic specimen – 'this marueilous straunge fishe' – was brought to London by the crew of Captain John Hawkins's second expedition, and exhibited with a hand-bill stating that: 'Ther is no proper name for it that I knowe, but that sertayne men of Captayne Haukinses doth call it a Sharke.'[11] Before then, sharks were known as 'dog fish' or 'hound fish' in several European languages, including English, though when Spanish merchants first encountered sharks in the Caribbean, they borrowed the Arawak word *tiburon* (Spanish *tiburón*), which briefly passed into English usage before being supplanted by 'shark'.

Why are sharks so widely feared? According to zoologist Edward O. Wilson, it's not fear we feel so much as fascination, with stories and fables woven around these creatures as a form of social defence, 'because fascination creates preparedness, and preparedness, survival: in a deeply tribal way, we love our monsters'.[12] In Hawaiian mythology, for example, there are stories of shark-men who change between shark and human form; sometimes a shark-man will warn beachgoers of sharks in the water, but the beachgoers laugh and ignore the advice, only to be attacked and eaten by the same shark-man who issued the warning. This is, in essence, the plot of *Jaws*, the suspensefulness of which – in both the novel and the film – was achieved through brief, unsettling glimpses of the 'spectral silver-grey blur' swimming silently under water, 'the jaw, slack and smiling, armed with row upon row of serrate triangles. And then the black, fathomless eye; the gills rippled – bloodless wounds in the steely skin.'[13] As predators, sharks depend upon elusiveness, and the many technical difficulties that beset the filming of the novel (not least the malfunctioning mechanical shark, nicknamed 'Bruce', after Spielberg's lawyer) ended up working to the film's advantage, with the shark itself appearing on screen for less than eight minutes of *Jaws*'s two-hour running time, and not appearing at all until eighty minutes in. Similar problems had plagued the filming of John Huston's over-budget adaptation of *Moby Dick* (1956), in which the main prop, a 23-metre (75 ft) rubber whale manufactured by Dunlop, came loose from its tow-line and was lost at sea off the Canary Islands. But Spielberg's decision to continue filming *Jaws* while his mechanical shark was out of commission proved the secret of the film's success, the suspense rising with every minute that the creature remained unseen.

Benchley's novel had a number of literary antecedents, most notably Melville's *Moby-Dick*, with Quint's obsession with the great white shark mirroring Captain Ahab's obsession with the great white whale. In the novel, Quint (captain of the *Orca*) drowns after being dragged underwater on a harpoon line, closely recalling the death of Ahab, who is pulled into the sea by the whale. But the infamous film version of Quint's death, in

'Bruce', the fault-prone mechanical shark, was the true star of Steven Spielberg's *Jaws* (1975), despite appearing on screen for less than eight minutes of the movie's two-hour running time.

which he is bitten in two by the shark as it rises onto the sinking boat, had its origins not in *Moby-Dick*, but in Edgar Allan Poe's only novel, *The Narrative of Arthur Gordon Pym of Nantucket* (1838), in which Pym and his shipmate Peters, adrift in the Southern Ocean, find themselves surrounded by aggressive predators:

> Toward evening saw several sharks, and were somewhat alarmed by the audacious manner in which an enormously large one approached us. At one time, a lurch throwing the deck very far beneath the water, the monster actually swam in upon us, floundering for some moments just over the companion-hatch, and striking Peters violently with his tail.[14]

The pair manage to escape the sharks' attentions, unlike Spielberg's doomed Quint, whose gruesome death underscored the film's relentlessly negative view of the creatures. So great was *Jaws*'s impact that a heightened fear of sharks led to a coastal killing spree in the 1970s, with shark populations falling drastically, especially along the eastern seaboard of the United

Winslow Homer, *The Gulf Stream,* 1899, oil on canvas. In this scene of imminent disaster, a man faces his likely death on a dismasted, rudderless fishing boat, threatened by hungry sharks as well as an approaching waterspout. The distant schooner on the horizon affords the only glimmer of hope.

States, where shark-killing competitions became a regular sport. Between 1986 and 2000, Northwest Atlantic hammerhead shark populations declined by 89 per cent, great white sharks by 79 per cent and tiger sharks by 65 per cent.[15] A quarter of all shark species, along with their cousins, the rays, currently face extinction. Benchley later wrote of his regret at the so-called '*Jaws* effect', and devoted the last decade of his life to marine conservation, publishing non-fiction books such as *Shark Trouble: True Stories About Sharks and the Sea* (2002), in which he advocated greater understanding and respect for the fish whose infamy he had done so much to establish. His amends-making advocacy work was posthumously recognized in 2015, when a newly identified species of lanternshark was named after him: *Etmopterus benchleyi*.

Sharks continue to face an uncertain future, with overfishing the greatest hazard. Over 100 million sharks are caught each year, mainly for their fins (the principal ingredient in shark-fin soup); usually, the dorsal and pectoral fins are sliced off and the rest of the animal thrown back into the sea, alive, the callousness compounded by humanity's atavistic fear of these ancient but now critically endangered creatures. How many of us, after all, can feel sympathy for a species born of intrauterine cannibalization, a pitiless reproductive strategy in which the strongest offspring devours its own embryonic siblings in the womb before entering the ocean alone? A finned shark, unable to swim effectively, will sink to the bottom of the sea and either die of suffocation or be eaten by other predators. The impact on shark populations is severe, particularly because sharks mature slowly, giving birth to fewer young than are needed to replenish their dwindling populations. If finning continues at its current levels, marine biologists predict that almost every shark species in our seas will be extinct by the end of this century.[16]

Curious creatures

In the decades since the release of *Jaws*, television documentaries have sought to enhance the public understanding of marine

Dumbo octopuses (*Grimpoteuthis* spp.) are elusive deep-sea foragers, named for their resemblance to Disney's baby elephant. They move by slowly flapping their prominent ear-like fins, while using their arms to steer.

environments, none more so, in recent years, than the BBC's land-mark *Blue Planet* series, which premiered in September 2001, with a follow-up series, *Blue Planet II*, released in 2017. The programmes reached global audiences on a level not seen since the pioneering American television series *The Undersea World of Jacques Cousteau* (1968–76), featuring a wealth of animal behaviour that had never previously been caught on film, such as cooperative hunting between bottlenose dolphins and false killer whales, or an orange-dotted tuskfish (*Choerodon anchorago*) using coral as an anvil on which to forcibly break open clam shells. Some of the filmed behaviour on the series was new to science, such as the footage of a common octopus clothing itself in shell debris in order to create a camouflaged hiding place from a predatory pyjama shark; but what stood out for viewers was the sheer strangeness of so much underwater life, from the extraordinary-looking Dumbo octopus (*Grimpoteuthis* spp.), which really does look like a cartoon character, to the hermaphroditic Asian sheepshead wrasse (*Semicossyphus reticulatus*), that can transition from female to male during the mating season, transforming its entire physiology within a matter of weeks.

The uncanniness of marine life increases with ocean depth. One of the strangest and most elusive of all sea creatures is the giant oarfish (*Regalecus glesne*), which can reach up to 11 m (36 ft) in length. First described in 1772, oarfish have rarely been seen alive, remaining as they do within the mesopelagic zone, between 200 m (660 ft) and 1,000 m (3,300 ft) down. These elongated, bony fish only rise to the surface when sick or dying, and have occasionally been found washed up dead on the shoreline, their large size and unsettling appearance provoking alarm and speculation. In July 2008 scientists used a remotely operated submersible to capture the first ever footage of a giant oarfish swimming in its natural deep-water habitat in the Gulf of Mexico. It was the first confirmed sighting of an oarfish at depth, with the animal estimated at between 5 and 10 m (16 and 33 ft) in length.[17]

Oarfish may well have been one of the origins of sea serpent stories told by early ocean travellers, and recounted in the natural histories of Aristotle and Pliny, the latter describing the

Engraving, from *Harper's Weekly*, of a 5-m-long oarfish that washed up on a Bermuda beach in 1860.

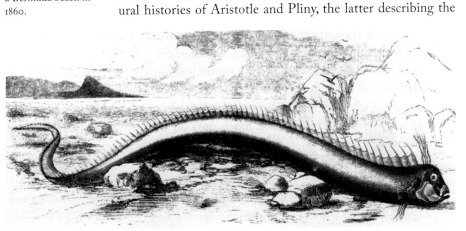

enormous skeleton of a 'sea monster' exhibited in Rome, which was '40 ft. long, the height of the ribs exceeding the elephants of India, and the spine being 1 ft. 6 inches thick'.[18] The skeleton had reportedly been washed ashore near Joppa in the province of Judaea, with Pliny arguing that it was the mortal remains of Ketos, the legendary sea monster sent by Poseidon to devour the captive princess Andromeda, before her dramatic rescue by Perseus. Legends aside, might the exacting eyewitness description of this creature point to it having been an ancient oarfish skeleton, or is it just a picturesque reminder of the continuity between extinction and myth, between outlandish creatures that nevertheless existed and those that never did except in the imagination?

Pliny's exhaustive (and at times, exhausting) *Natural History* is characterized by a curious combination of research and gossip, mixing sober observational testimony with fantastical stories, such as the tale of the dolphin who fell in love with a human boy and pined to death when the boy moved away, or the eyewitness account of a merman, a 'man of the sea', that appeared in Book IX:

Sea lions resting on a shipping buoy in the waters of southeast Alaska.

I have distinguished members of the Order of Knighthood as authorities for the statement that a man of the sea has been seen by them in the Gulf of Cadiz, with complete resemblance to a human being in every part of his body, and that he climbs on board ships during the hours of the night and the side of the vessel that he sits on is at once weighed down, and if he stays there longer actually goes below the water.[19]

What was this unearthly creature and was it related to the various semi-divine sea-spirits that Pliny also described, from the Triton seen playing a conch shell in a sea cave to the human-shaped 'Nereids' that populated the Atlantic shoreline, 'whose mournful song moreover when dying has been heard a long way off by the coast-dwellers'?[20] Were these seals, perhaps, or a colony of stranded sea lions?

Pliny's account also included an early description of a giant squid (or 'polypus'), with a head the size of a barrel, and vast tentacles with which it attacked and capsized fishing boats. According to Erik Pontoppidan (the Lutheran bishop of Bergen) in his *Natural History of Norway* (1752), this was the legendary kraken, 'a marine animal of such enormous size that it more resembled an island than an organised being'.[21] Like Pliny with his mythical 'Ketos', Pontoppidan outlined numerous instances of kraken sightings across the North Atlantic, including the testimony of Norwegian fishermen who had seen 'this enormous monster come up to the surface of the water', although its entire body is never seen at once, only 'what looks at first like a number of small islands surrounded with something that floats and fluctuates like sea-weeds':

at last several bright points or horns appear, which grow thicker the higher they rise above the surface of the water, and sometimes they stand up as high and as large as the masts of middle-sized vessels. It seems these are the creature's arms, and it is said if they were to lay hold of the largest man of war they would pull it down to the bottom.

After this monster has been on the surface of the water a
short time it begins slowly to sink again, and then the danger is as great as before; because the motion of his sinking
causes such a swell in the sea, and such an eddy or whirlpool,
that it draws everything down with it, like the current of the
river Male.[22]

Along with colossal squid (*Mesonychoteuthis hamiltoni*) and
gigantic octopi (*Octopus giganteus*) – two recognized, albeit elusive species of giant cephalopod – one of the many candidates
for the origins of the kraken legend was undersea volcanism in
the waters off Iceland, which can indeed perturb the sea with
sudden, hazardous currents, or break the surface with the tops
of new islets. This idea was hinted at in Alfred, Lord Tennyson's
irregular sonnet 'The Kraken' (1830), in which new scientific
theories about the age and fiery origins of the earth were combined with ancient Scandinavian lore:

Below the thunders of the upper deep;
Far, far beneath in the abysmal sea,
His ancient, dreamless, uninvaded sleep
The Kraken sleepeth: faintest sunlights flee
About his shadowy sides: above him swell
Huge sponges of millennial growth and height;
And far away into the sickly light,
From many a wondrous and secret cell
Unnumbered and enormous polypi
Winnow with giant arms the slumbering green.
There hath he lain for ages and will lie
Battening upon huge sea-worms in his sleep,
Until the latter fire shall heat the deep;
Then once by man and angels to be seen,
In roaring he shall rise and on the surface die.[23]

For Tennyson, the rising of the kraken symbolized some
tumultuous future apocalypse, either human or geological, that
would erupt from the unfathomable depths, from this 'bizarre

'In roaring he shall rise and on the surface die': a giant squid attacking a boat at sea, as imagined in an early 19th-century engraving.

marine version of the Tennysonian garden', as literary critic Seamus Perry has characterized it.[24] Powerful undersea giants went on to become a staple of nineteenth-century adventure stories. In H. G. Wells's unsettling tale 'The Sea Raiders' (1896), a pod of giant flesh-eating squid, 'mysteriously arisen from the sunless depths of the middle seas', launches a series of deadly attacks along the Devon coast. The story adopts the tone and terminology of a scientific paper ('in the case of *Haploteuthis ferox*, for instance, we are still altogether ignorant of its habitat'), but soon descends into full horror mode, with the first sighting of a group of squid devouring a human body on the shoreline, and the gruesome death of a member of a boat party sent out to investigate 'these abominable creatures':

> He lifted his arm, indeed, clean out of the water. Hanging to it was a complicated tangle of brown ropes, and the eyes of one of the brutes that had hold of him, glaring straight and resolute, showed momentarily above the surface. The boat heeled more and more, and the green-brown water came pouring in a cascade over the side. Then Hill slipped and fell with his ribs across the side, and his arm and the mass of tentacles about it splashed back into the water . . . and in another moment fresh tentacles had whipped about his waist and neck, and after a brief, convulsive struggle, in which the boat was nearly capsized, Hill was lugged overboard.[25]

Wells's story recalled a range of ancient sea monsters, including the kraken, and the six-headed Scylla of Greek mythology, as well as more recent literary encounters with giant marine creatures, such as Herman Melville's *Moby-Dick*, or Jules Verne's *Twenty Thousand Leagues Under the Seas*, one of the most dramatic sequences of which, the battle against a school of giant squid, begins when a crewman opens the hatch of the boat and gets caught by one of the 9-metre (30 ft) monster's tentacles: 'immediately one of these long arms glided like a serpent through the opening, and twenty others were brandished above it. With a blow of the hatchet Captain Nemo cut off this formidable tentacle, which glided twisting down the steps.'[26]

It has often been pointed out that Verne gave his squid eight arms, like an octopus, whereas in fact they have ten, but given that the giant squid had not yet been recognized by science, it was an understandable mistake. To add to the ambiguity, Verne did not use the word *calamar* (French for 'squid'), but *poulpe*, which can also be translated as 'octopus'. And at the height of the underwater fight scene, Verne's narrator, Professor Aronnax,

A giant squid washed ashore at Trinity Bay, Newfoundland, in 1877. Engraving published in *Canadian Illustrated News*, October 1877.

observes that to do justice to the attack would require 'the pen of the most illustrious of our poets, the author of "The Toilers of the Deep"', yet another work of nineteenth-century fiction that featured a long description of a sea-monster attack, in this case, a giant octopus.[27] The scene appeared in Victor Hugo's great maritime tale *Toilers of the Sea* (1866), in which the novel's hero Gilliatt fights to the death with a fearsome 'devil-fish' (or giant octopus), which is described with hallucinatory clarity:

> This irregular mass advances slowly towards you. Suddenly it opens, and eight radii issue abruptly from around a face with two eyes. These radii are alive: their undulation is like lambent flames . . . A terrible expansion! . . . Its folds strangle, its contact paralyses. It has an aspect like gangrened or scabrous flesh. It is a monstrous embodiment of disease . . . Underneath each of [its] feelers range two rows of pustules, decreasing in size . . . They are cartilaginous substances, cylindrical,

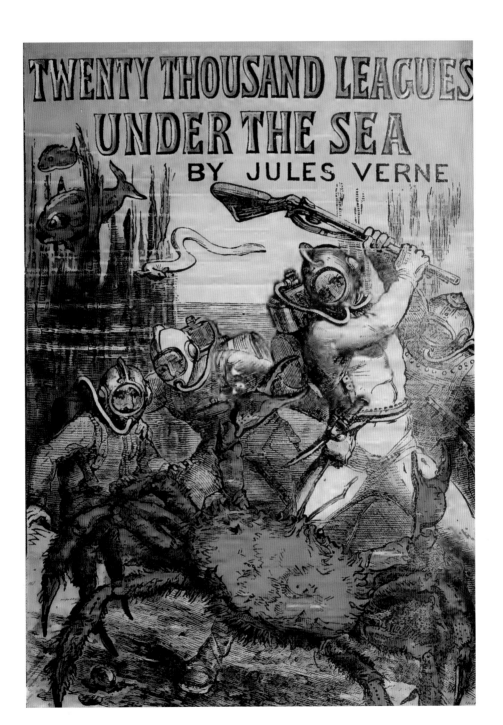

TWENTY THOUSAND LEAGUES
UNDER THE SEA
BY JULES VERNE

The crew of the *Nautilus* attacked by a giant deep-sea crab, as depicted on the cover of a popular English-language edition of Jules Verne's *Vingt mille lieues sous les mers*.

horny and livid ... The devil-fish is crafty. When its victim is unsuspicious, it opens suddenly. A glutinous mass, endowed with a malignant will, what can be more horrible?[28]

In contrast to the aquadynamic squid, the octopus is a boneless bag of soft tissue with no fixed shape or colour, which is able to wriggle and slide over any obstacle it meets, making its way through the tiniest crevice. Even the largest octopus species, the Giant Pacific (*Enteroctopus dofleini*), can squeeze through a 2-centimetre opening in spite of its 5-metre (16½ ft) arm span, which is why they have proved difficult to keep in captivity. Octopuses encountering divers in the wild will often greet them with an inquisitive arm, and sometimes even shake them by the hand. Aristotle, in his *Historia animalium*, called the octopus (or *polypous*) 'a stupid creature' for this handshaking habit, which has historically made it one of the easiest sea creatures to hunt, but in fact these three-hearted, eight-armed molluscs are remarkably intelligent. They are masters of disguise, able to change their skin colour at will, and can camouflage themselves into virtual invisibility as a defensive tactic (seals, sea otters, sharks and whales are regular predators) as well as squirt clouds of ink and mucus to provide cover for a quick escape. Particularly impressive is their 'moving rock' trick, which involves the octopus mimicking an undersea rock before inching across the open sea floor at a speed that matches the movement of the water, allowing the creature to move in plain sight of a predator.

The word *octopus* ('eight-footed') was first used in print in Guillaume Rondelet's *De piscibus marinus* (1554), in which the naturalist referred to the *poulpe commun* (common poulpe) by an alternative 'scientific' name, *Polypus octopus*, which was later taken up by the Swedish botanist Carl Linnaeus in his founding work of classification, the *Systema Naturae* (10th edn, 1758), establishing 'octopus' as the genus and 'octopodia' (later 'octopoda') as the order of cephalopod molluscs with eight sucker-bearing arms. Linnaeus described the octopus as a *Singulare monstrum*, 'a unique monster', a creature that both fascinates and repulses; 'partly fish, partly reptile', as Victor

Hugo characterized it a century later, 'soft and flabby; a skin with nothing inside. Its eight tentacles may be turned inside out, like the fingers of a glove.'[29]

Whales

In 1989, scientists at the Woods Hole Oceanographic Institution in Massachusetts detected the song patterns of a migrating whale that were pitched at the unusually high frequency of 52 hertz, a frequency at which no other migratory whale has been known to sing. Blue whales, for example, sing at between 10 and 39 hertz, while most fin whales sing at around 20 hertz. Over the years, the mystery whale's calls have deepened slightly to around 49 hertz, as the unseen creature matured and grew, but it continues to broadcast a uniquely high-pitched song. It has been detected in the Pacific Ocean every year from around August to December, travelling as far north as the Aleutian and Kodiak Islands, and as far south as the California coast, swimming between 30 and 70 km (18–43 mi.) every day. The 52-hertz whale, nicknamed 'the loneliest whale in the world', has roamed the Pacific for at least the past thirty years, calling futilely for a mate at a frequency no other whale can hear.

It's a sad story, but the high degree of public interest in the plight of the lonely whale shows how much human attitudes to these creatures have changed. Until the 1960s, whales were still being hunted in enormous numbers, including the blue whale, the largest animal ever to have existed, which was driven to near-extinction until the International Whaling Commission (IWC) banned blue whale hunting in 1966. Whaling remains an emotive subject, and despite the more general moratorium on commercial whaling introduced twenty years later, in 1986, a handful of countries continue to allow the hunting of whales, including Canada, Iceland, Japan, Norway, Russia, South Korea and the United States, as well as the Danish dependencies of the Faroe Islands and Greenland. Currently, more than 1,500 whales are caught and killed every year, both for culinary and scientific purposes, although that number could well be higher: memoirs of Russian whaling inspectors published after the 1986 moratorium revealed that between 1959 and 1961, Soviet whaling fleets killed at least 25,000 humpback whales in the Southern Ocean, while reporting a catch of just 2,710, and the official IWC figures for Japanese whaling may well have been equally under-reported.[30]

Coastal peoples have always made use of whales, whether caught or washed up onto the shore, and a number of North Sea communities were once dependent on the animal for food. In George Mackay Brown's poem 'The Year of the Whale' (1965), a village of starving Orcadians, having subsisted through the winter on 'limpets and crows', spot a school of whales off Scabra Head, and take to their boats armed with hoes and ploughshares to drive the 'wallowing lumps of thunder and night' ashore:

Then whale by whale
Blundering on the rock with its red stain
Crammed our winter cupboards with oil and meat.[31]

The improvised nature of the Orcadian hunt suggests an opportunistic approach, in contrast to the systematic hunting of whales that developed elsewhere, beginning with the vast schools of North Atlantic right whale, *Eubalaena glacialis* ('the true whale of ice'), which was prosaically named the 'right whale' because it was, quite simply, the right whale to hunt, being slow moving and docile, feeding near the surface with its mouth wide open, filtering krill through its baleen plates. Unfortunately for right whales, they float after death due to their high blubber content (unlike the heavier blue and fin whales), making them easy to retrieve from the water. Their flexible baleen plates, which are formed from keratin but were marketed as 'whalebone', were prized in the manufacture of a range of domestic items, from corsets to umbrellas, while their plentiful blubber was rendered into whale oil, which, in the years before domestic gas supplies, proved the cheapest and most abundant source of lighting. By the 1740s London alone had 5,000 street lamps fuelled by whale oil, and as new technologies appeared, the application of whale products expanded to include machine oil, soap and margarine. Sperm whales also secrete a waxy substance known as ambergris, or whale amber, which became a valuable ingredient in the perfume industry: as Herman Melville noted in *Moby-Dick*, a chapter of which is devoted to the quest for ambergris, there was an irony in the fact that 'fine ladies and gentlemen should regale themselves with an essence found in the inglorious bowels of a sick whale!'[32]

Ships flocked to northern waters in pursuit of whales; by the late 1780s, some 250 British whaling ships were at work in the Arctic, and Norway had become the centre of the world's whaling industry. In 1863 the invention of the modern harpoon gun, which used steam power and an explosive tip, designed to deliver a fatal impact, transformed the whaling industry. The new harpoon allowed whalers to move from hunting right whales to hunting wrong ones – the blue and fin whales, faster swimmers that sank when killed, and so had remained untouched while northern right whales were being hunted to the point of extinction. A new phase of hunting began, with fin whale numbers soon plummeting to the level of the now all-but-extinct right whale. As the doomed ship's commander, Brownlee, in Ian McGuire's 1860s-set whaling novel *The North Water* (2016) observed, 'twenty years ago, the waters about here were full of whales, but they've all moved north now – away from the harpoon. Who can blame them? . . . steam is the future, of course. With a powerful enough steam ship, we could hunt them to the ends of the earth.'[33]

And Brownlee was right. Whales might be intelligent, cultural animals that are able to pass on knowledge of risk, but they are no match for explosive harpoons. Whaling only made economic sense where whales were abundant, and so whaling activity shifted south, towards the Antarctic. Ships from Norway and Scotland ventured into the Southern Ocean, and reported that there were hundreds of thousands of blue, fin and humpback whales thriving in the Antarctic Convergence, the point where relatively warm Atlantic water meets that of the cold Antarctic: where the two bodies of water meet, the associated turbulence and upwelling stirs up nutrients from the seabed,

A Dutch whaling crew approach a bowhead whale in the icy Arctic waters off Spitsbergen, as depicted in an oil painting by Abraham Storck (1690). Bowhead whales (also known as Greenland right whales) were an early target of Arctic whaling, and despite the 1966 moratorium, remain on the endangered list.

supporting a surprising abundance of marine life, from krill upwards, including vast numbers of whales. South Georgia soon overtook Norway as the centre of the whaling industry, with factory ships taking thousands of whales each year, mostly humpbacks (*Megaptera novaeangliae*) and southern right whales (*Eubalaena australis*), which proved so trusting and inquisitive that they swam right up to the whalers' boats. Such were the numbers that whalers often took only the blubber from their catches, leaving the carcasses to sink to the seabed, whalebone by then having fallen out of fashion and demand.

By 1914 it was clear that the depletion of whales in northern waters was being replicated in the south. The British government set up a committee to investigate, despatching Ernest Shackleton to South Georgia to compile a report, but the outbreak of war in Europe created a renewed surge in demand for whale oil in nitroglycerine production, and the nascent restrictions on whaling were dropped. The years after the First World War saw a further rise in demand for whale oil as a cheap alternative to other animal fats in cooking and soap production, as well as for bonemeal fertilizer rendered from the vast skeletons. In a gruesome irony, whales were now being killed by explosive harpoons tipped with nitroglycerine made from their own blubber. The Edinburgh-based whaling company Christian Salvesen, then the largest commercial presence in South Georgia, hit peak productivity in 1925, when 16,000 tonnes of whale oil were sent back to the UK for a profit worth more than £30 million today. By then, the whaling process was conducted on an industrial scale: a 90-tonne whale could be flensed and butchered in twenty minutes; in 1926 the largest living animal ever recorded on Earth, a 33-metre-long (108 ft) blue whale, many decades old, was killed, dragged onto the flensing platform and dismembered in less than an hour. But while the Colonial Office in London welcomed the tax revenue from the Antarctic whaling boom, it also recognized that industrial whaling stations were proving too effective and that Southern Ocean whale stocks were disappearing fast.

When the price of whale oil collapsed in the 1950s, following the introduction of cheaper vegetable oils in soaps and cooking, Salvesen sold its whaling quota to the Japanese, ending half a century of British whaling activity in the Antarctic Ocean, during which more than 1.5 million whales had been caught and killed, and the global whale population brought to the brink of extinction.

The hunting bans of the 1960s reflected a growing conservationist awareness that was fuelled by increased understanding and respect for these impressive marine mammals. Like humans, whales sing, or rather vocalize: male humpback whales compose individual musical pieces as a means of communicating with females, while the sperm whale's head features nature's largest sound-producing organ, which can weigh up to 11 tonnes. The clicks it makes have been measured up to 230 decibels, and can travel for enormous distances through water, through which the speed of sound is roughly four times greater than in air.

As this woodcut from André Thevet's *Cosmographie universelle* (1575) graphically shows, flensing and butchering a whale carcass was a complex and bloody undertaking.

Whale behaviour contains a world of surprises. Diving sperm whales, for instance, prey on squid and octopus, grabbing a tentacle or two on the way down before crushing the animal to death against the sea floor. At the bottom, the whale might swim upside down, scanning the water above for prey silhouetted against the faint light above, with sperm whales deploying sonar to identify schools of fish or squid which it then swims back up to catch. Baleen whales, such as the blue whale, by contrast, feed almost solely upon krill, prolific shrimp-like crustaceans that measure less than 6 cm (2⅓ in.) long. A blue whale can consume up to 4 tonnes of krill per day by taking in huge mouthfuls of krill-rich water; it then strains the water through its baleen plates, retaining the nutritious food. Despite their gargantuan appetites, whales help sustain the diversity of marine life; by eating at depth and excreting at the surface, they recycle nutrients from the depths, a phenomenon known as the 'whale pump'.[34] And when they die, their carcasses sink to the ocean floor, delivering nutrients back into the marine ecosystem.

The Southern Ocean is one of the richest on Earth, with strong currents stirring up nutrients from deep water, creating the most abundant of all marine feeding grounds, based on vast shoals of Antarctic krill (*Euphausia superba*) – an estimated 400 trillion of them – with a combined weight greater than any other animal species on the planet.[35] Krill are a keystone prey species, upon which almost every living thing in the Southern Ocean depends. Whales, penguins, seabirds, squid, seals and fish all feed directly on these small crustaceans – one of the principal sources of protein on the planet – but in recent decades, krill populations have been in steep decline, while also moving towards the poles in response to rising sea temperatures. During winter, krill migrate into deeper water, over which sea ice spreads, trapping large quantities of algae beneath the ice. When the ice melts in the spring, the algae blooms, and the newly hatched krill rise to feed upon it (and are themselves fed upon by other creatures). But sea ice is now melting earlier than ever, and if the young krill rise earlier to feed, they are in a slow metabolic state, and feed inefficiently, resulting in the sub-ice algae dying and sinking to the sea floor before the krill have eaten their fill. Such phenological mismatches, also known as trophic asynchronies, occur when interacting species change the timing of long-established phases in their life cycles at different rates, and can lead to catastrophic effects. The impacts of climate change on krill, particularly a reduction in juvenile populations, have been observed in both Arctic and Antarctic contexts, and as a research paper published in 2019 concluded, if krill distribution trends continue as they are, a damaging population slump confronts every species that relies upon them, from seabirds to whales.[36] There is also a significant commercial fishing sector, which harvests around 200,000 tonnes of krill each year, mainly for fish food and bait, but also for a growing global market in omega-3-rich health products, which places further stress on krill populations. Some Southern Ocean species, including fur seals and macaroni penguins, are already finding it harder to source enough krill to support their populations, as evidenced by the falling birth weights recorded by seal researchers on South Georgia and elsewhere.

Adult krill populations have fallen by nearly 80 per cent since the 1970s: a shocking statistic that ought to be headline news, for while the looming krill crisis may be less familiar and less emotive than many other environmental risk stories, it places the future of a vast array of marine species in real jeopardy. As polar temperatures continue to rise, the fate of this humble crustacean may prove to be one of the main determining factors in the future life of our seas.

4 Exploration

The sea is no place to be if you can help it, and to try to cross it
betrays a rashness bordering on hubris.
W. H. Auden, *The Enchafèd Flood* (1950)

One of the most remarkable objects in the National Maritime
Museum in Greenwich is a flimsy-looking reed lattice, studded
with shells, that – on closer inspection – reveals itself to be a
sophisticated visual encoding of the waves and currents encountered in the sea around the Marshall Islands in the western
Pacific. The islands on this traditional 'stick chart' are represented
by cowry shells, with the swells and movements of the sea around
them indicated by the varying patterns of the reeds. The curves
show where swells are deflected by an island; the short, straight
strips indicate the currents flowing near land, while the longer
strips are thought to indicate compass direction. These frail
physical objects, unique to Micronesia, seem to have been used
as mnemonic aids on shore rather than charts for use at sea, but
they illustrate an expert understanding of the complex interactions of land and water that made the argonauts of the western
Pacific such exceptional navigators.[1]

The epic sea voyages of early Oceania remain the greatest
feats of navigation in human history. The first canoe journeys
were made more than 50,000 years ago by descendants of Ice
Age hunters in Southeast Asia, who migrated over the South-
western Pacific to populate New Guinea, Australia and the
islands of the Bismarck Strait. These large islands, many visible
from one to the next, would have provided a voyaging corridor
from mainland Asia to the end of the Solomons in Near Oceania, with the alternating northwest monsoon and southeast
trade winds making regular return journeys possible. This first

Micronesian 'stick chart', on which islands are marked by cowry shells, with the swells and movements of the sea around them indicated by the patterns of the latticed framework.

phase of Pacific settlement was completed around 25,000 years ago, at which point the voyaging urge appears to have waned until around 3,000 BC, when a new seafaring civilization, known to archaeologists as the Lapita culture, developed the maritime skills that saw them travel from island to island as far east as Fiji, on the threshold of Polynesia. This second phase of inter-island exploration and settlement seemed to pause again for several centuries, but in around AD 900 long-distance voyaging resumed, and within two hundred years, descendants of the earlier Lapita people had made their way to Polynesia's Cook and Society Islands, more than 3,000 km (1,900 mi.) to the east of Fiji. In around AD 1100 Polynesian landfinders began voyaging southwest in double-hulled canoes, reaching the distant islands of Aotearoa (New Zealand) sometime in the thirteenth century. And some may even have sailed further still, as far as the Pacific coast of South America, centuries before European navigators such as Magellan and Cook made their way into the heart of Polynesia on well-appointed carracks and colliers.

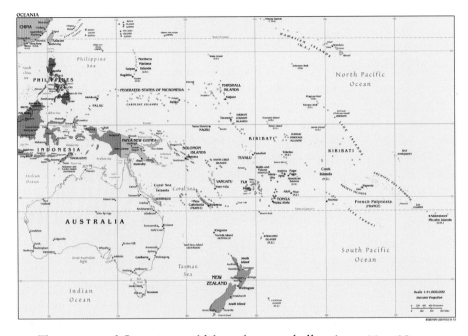

The waters of Oceania would have been a challenging environment for these early voyagers, who faced vast horizons of deep ocean with no land in sight, and journeys of many hundreds, sometimes thousands, of kilometres between groups of unknown islands. It can only be guessed at why those pioneers took to the sea in vulnerable vessels, on logs and rafts and dugout canoes. Was it hunger for food or land, war or territorial conflict, a search for prestige through trade and exchange, or 'simple restlessness and curiosity about what lay beyond the blue horizon', as Brian Fagan has phrased it?[2]

Map of Oceania. Early navigators voyaged enormous distances between far-flung islands in the 8.5 million sq. km of Oceania's waters.

The navigational techniques of the Pacific argonauts also remain a matter of debate, and at times a degree of scepticism has been directed at the historical accuracy of the epic voyaging narratives that are central to the stories of Polynesian navigation. The New Zealand historian Andrew Sharp, for example, writing in the 1950s, argued that planned two-way voyages without instruments were impossible over such long distances, and that the more remote Pacific settlements, such as Easter Island, and New Zealand itself, must have been found accidentally by one-way drift voyagers travelling with no intention of returning home.[3] He also disputed the seaworthiness of the Islanders' canoes, arguing that most were too cumbersome to be pointed accurately into the wind. However, the notion of Pacific islands as a series of tiny specks separated by an immense blue void may well be a Western construct, one pre-dated by a more holistic understanding of the ocean as an inhabited realm that unites a large exchange community; this alternative perspective was supported by later work by anthropologists such as David Lewis, who used reconstructed technology manned by local seafarers, who demonstrated that traditional navigation methods were every bit as accurate as modern approaches. Lewis, who sailed more than 20,000 km (12,500 mi.) through Pacific waters under the instruction of island navigators, reported that these skilled landfinders followed star paths, oriented by the Sun, and used dead reckoning, much as mariners do today, employing deep navigational knowledge transmitted down the generations through songs and stories that told of ocean currents, bird routes, whale paths and wind patterns.

Of course these time-honoured techniques could only be used when the voyagers knew both where they were and in which direction they were going. If they were exploring a new area of ocean, or had been blown off course by a storm, they turned to other methods, such as observing the flights of frigate birds, or the dispositions of cumulus clouds, which tend to dip in V-formations down towards an island, or be locked onto it while other clouds pass by over the water (hence the Maori name for New Zealand: Aotearoa, 'the land of the long white cloud').

But the system most favoured by Polynesian and Micronesian navigators involved steering by ocean swells. Swells are waves that have travelled beyond the winds that produced them, and even when flattened and distorted by local air movements, they remained interpretable by Pacific navigators, who steered by the feel of the waves beneath their canoes. David Lewis described how trainee navigators in the Marshall Islands were taken out to sea and instructed to lie flat in the water:

These elder skippers, first of all, would take the younger man out to the ocean. They would be in a boat, but they would lay the young man in the water, on his back, and tell him to float and relax so that he would get to know the feel of the waves as they came along.[4]

Such was their deep somatic connection to the sea that older Marshallese navigators often chose to lie prone in their canoes, pressing an ear to the boards for minutes on end before calling out directions to the helmsmen, while others stood, carefully balanced, 'plumbing the swell', feeling its effect through their bodies in subtle shifts from the vertical.[5] Lewis recalled an episode in which a Gilbert Islands navigator named Iotiebata was blown off course by a northwest storm that came at the end of a month of westerly gales; but he was still able to detect through his body the underlying easterly swell that was 'sweeping by unbroken by any land. Such swells continue to come from the east regardless of the direction of the wind, even a storm wind.'[6]

The mechanics of ocean swells are complex, yet ultimately learnable over the long years of training. When a swell meets a reef or an island, part of it bounces directly off, forming reflected waves of shorter length, while part of it is refracted, dragging where it touches the shore, and creating turbulent cross-currents and distortions. If an island deflects a prevailing swell, its presence can be 'felt' by a sensitive navigator, even from a considerable distance. It was these kinds of interference patterns that the Micronesian stick charts were designed to record, with some of them – the so-called *rebbilib* charts – showing known routes and currents by swell, and others – the *mattang* charts – illustrating more general principles of wave interference (such as the way a swell steepens as sea depth decreases, heralding an approach to land), which could be used as keys to understand new and unfamiliar waters.

By these means the early Pacific navigators voyaged over time to every island within the vastness of Oceania – some of the remotest specks of habitable land on Earth – their only technologies being the long, double-hulled canoe with its claw-shaped sail, and an intimate, centuries-won knowledge and understanding of the swells and circuits of the sea.

Voyages of empire

As the islands of Oceania were being colonized and populated by trans-Pacific navigators, western trade routes were also beginning to be established across the Mediterranean and beyond. By 1800 BC the Minoans were in control of a Bronze Age maritime empire based on Crete, from where they traded in oil, grain, wine and other commodities, including ideas, giving rise to the world's first commercial civilization along the coastlines of the eastern Mediterranean. Their dominance lasted until circa 1450 BC, when a catastrophic volcanic eruption brought an end to Minoan culture, opening up the seas instead to the neighbouring Phoenicians, who went on to build a powerful trading culture, a thalassocracy ('seaborne empire') based on new technologies of sail.

Trade is a two-way process, with trading by sea requiring winds blowing in both directions, 'or at least ships manoeuvrable enough to make it seem as though they do,' in Lyall Watson's words.[7] While the ceremonial long boats of other Mediterranean nations developed into cumbersome galleys or military triremes, the Phoenicians developed smaller, practical merchant sailboats that could tack into the wind along the coastlines. By 1200 BC the coast of Syria was studded with Phoenician cities connected by regular sailing routes to Cyprus, the Greek mainland, Sicily and Spain; within a few centuries, they had extended their routes along the North African coast, through the Strait of Gibraltar and out into the Atlantic, circumnavigating the continent of Africa, according to Herodotus, who described the epic journey of 'a fleet manned by a Phoenician crew with orders to sail west-about and return to Egypt and the Mediterranean by way of the Straits of Gibraltar'.[8]

The Phoenicians would later make their way into the Indian Ocean, although they were not the first voyagers to do so. As the Norwegian ethnographer Thor Heyerdahl pointed out, the

hieroglyphic sign for 'ship' in the earliest-known Sumerian script was identical to the ideogram for 'marine' in ancient Egypt; both showed the outline of a sickle-shaped reed boat with a high bow and stern, a design that Heyerdahl argued bore no resemblance to flat-bottomed river craft, but was clearly an adaptation to 'navigation through surf and big seas' beyond the mouth of the Nile.⁹ He went on to quote the polymath Eratosthenes, chief librarian of Alexandria during the early second century BC, who described how 'papyrus ships, with the same sails and rigging as on the Nile sailed as far as Ceylon (Sri Lanka) and further on to the mouth of the Ganges', as evidence that Egyptian traders had already made their way to India and back, aided by the seasonal reversal of wind and current directions across the monsoon area.¹⁰

Phoenician trading ship riding the waves, engraved on a 2nd-century sarcophagus, Sidon, Lebanon.

As seafarers had known since prehistory, the equatorial trade winds (their name derives not from commerce, but from the late Middle English word for 'path' or 'track') were excellent engines for one-way traffic across the Indian Ocean, but return journeys depended upon the monsoon, a reliable reverse trade wind that allowed for the two-way commercial sailing that so enriched this part of the world, and still does:

Between November and March, fleets of outbound dhows still sail from the Gulf, Pakistan and India to Mogadishu, Lamu, Mombasa and Dar es Salaam, flicking out at the tail of the wind to the Comores and Malagasy, carrying cargoes of spices, carpets and salt fish for sale in African bazaars. In April they rest and sleep and patch their sails, waiting for spring in the north and a time when Asia warms enough to draw the southeast trades once again out of the lower Indian Ocean.[11]

By May and June, the warm monsoon will have advanced over southern Asia, and the dhows, laden with cargoes of charcoal and grain, pick up the gathering winds and make the reverse journey home, in a long-established schedule dating back to the Bronze Age. Once the workings of the monsoon had been understood by later European traders, east–west traffic increased enormously, and by the time Claudius became emperor in AD 41, seaborne trade with India was booming. In around AD 60, the Graeco-Roman *Periplus of the Erythraean Sea* appeared, a comprehensive guidebook to the coastlines around the Indian Ocean ('Erythraean Sea' translates as 'Red Sea', by which the Greeks referred to both the Indian Ocean and the Persian Gulf). The *Periplus* detailed the region's ports, anchorages, markets, tides and sailing itineraries, from Egypt to India, in 66 vividly written chapters that were clearly the work of an experienced traveller who knew the territories well. The entry on Barbarikon (a trading port near the modern-day city of Karachi, Pakistan, on 'the seaboard of Scythia'), gives a flavour of the guidebook's detailed coverage:

The ships lie at anchor at Barbaricum, but all their cargoes are carried up to the metropolis by the river, to the King. There are imported into this market a great deal of thin clothing, and a little spurious; figured linens, topaz, coral, storax, frankincense, vessels of glass, silver and gold plate, and a little wine. On the other hand there are exported costus, bdellium, lycium, nard, turquoise, lapis lazuli, Seric skins,

Alexander Fights a Sea Battle, folio from an illustrated manuscript of a Khamsa (Quintet) of the Indian poet Amir Khusrau Dihlavi, 1597–8.

cotton cloth, silk yarn, and indigo. And sailors set out thither with the Indian Etesian winds, about the month of July, that is Epiphi: it is more dangerous then, but through these winds the voyage is more direct, and sooner completed.[12]

Bigger and more seaworthy ships were financed and built specifically for the expanding Indian Ocean trade, which led to equivalent advances in Mediterranean shipping. A second-century Tamil poem invoked 'the fine large ships of Yavana [Greeks] bearing gold, making the water white with foam', while the Syrian satirist Lucian, in his *Navigium*, described an enormous merchant ship '180 feet long, 45 feet wide, 44 feet deep, with a crew like an army, passengers of both sexes, and corn sufficient to supply Attica for a year', which reportedly delivered a large annual profit to its Egyptian-Greek owner.[13] Trading by sea may have been hazardous, but it was a fast route to enormous wealth. As Pliny the Elder observed in his *Natural History*, 'what is there more unruly than the sea, with its winds, its tornadoes, and its tempests? And yet in what department of her works has Nature been more seconded by the ingenuity of man than in this, by his inventions of sails and of oars?'[14]

Following the collapse of the Roman Empire, however, European contact with Asia diminished, only picking up again in the late fifteenth century, after Vasco da Gama commanded the first recorded voyage direct from Europe to India via the Atlantic, more than 2,000 years after Phoenician navigators had made their own circuitous way around the coast of Africa.

Christopher Columbus's unsuccessful search for a western maritime route to India resulted in the so-called 'discovery' of the Americas in 1492, but it was da Gama who ultimately established the *Carreira da Índia* (India route), when he sailed around Africa and into the Indian Ocean, landing at Calicut (modern-day Kozhikode) in May 1498. The expedition came at a heavy cost – two ships and over half the crew were lost – and it also failed in its principal mission of securing a commercial treaty with Calicut. Nonetheless, the small quantities of spices and other trade goods brought back on the remaining ships demonstrated the

The Portuguese explorer Vasco da Gama lands at Calicut, India, on 20 May 1498, as depicted in a late 19th-century painting by Ernesto Casanova. This was the first recorded sea voyage made directly from Europe to India via the Atlantic.

potential profit for future trade, and within a few years of da Gama's expedition, the Portuguese had established a seaborne trading empire through the military conquest of the strategic ports of Goa (in 1510), Malacca (in 1511) and Ormuz (in 1515). All the most important sea routes and trading networks of the Indian Ocean, Persian Gulf and South China Seas were soon under the control of the Portuguese crown, which had ruthlessly re-established a model of empire built on control of the seas, aided by state-funded technical developments in nautical science, especially in the field of cartography.

The so-called 'golden age' of Portuguese navigation was crowned by Ferdinand Magellan's expedition of 1519–22, which resulted in the first circumnavigation of the earth, although Magellan's great voyage had in fact been commissioned by the Spanish king, Charles I, rather than the Portuguese monarch, Manuel I, who had refused Magellan's petitions to fund a west-ward expedition to the Maluku Islands (the fabled Indonesian 'Spice Islands') avoiding the southern tip of Africa. With the king of Spain's blessing, Magellan's fleet of five vessels – the *Armada de Molucca* – headed southwest across the Atlantic to the coast of Patagonia. Despite a series of storms and mutinies, the ships eventually made it through what is now known as the Strait of Magellan into a body of water that Magellan named *Mare pacificum*, the 'peaceful sea', thereby naming the Pacific Ocean. The expedition slowly continued west to the Philippine islands, where Magellan was fated to be killed in battle in April 1521. Seven months later, the remaining two ships of fleet, now commanded by the Spanish navigator Juan Sebastián Elcano, reached the Indonesian Spice Islands. The voyage had answered its purpose, identifying a new western trading route to the Moluccas (the Maluku Islands), since the eastern route was already controlled by the Portuguese. But the new itinerary was not commercially viable, since crossing the Pacific had proved far longer and more arduous than any European could have expected. Magellan, anticipating 'a short cruise to the Spice Islands, followed by a longer but untroubled voyage home through familiar waters', could have had no conception of the

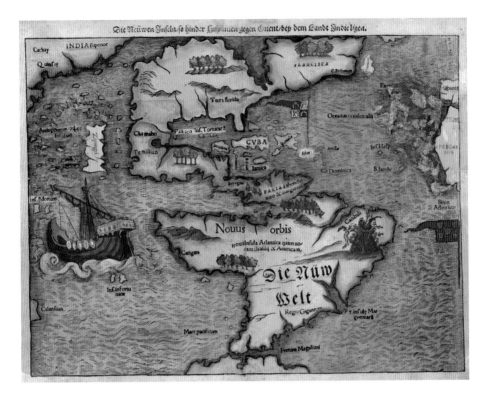

true scale of the Pacific, which encompasses one-third of our planet's surface, extending over a greater area than all the dry land on Earth.[15] If Magellan's voyage proved anything, it was the unthinkable scale of the ocean that he had named only months before his untimely death. The following year, in September 1522, his armada's sole surviving ship, the *Victoria*, returned home to Spain via the Indian Ocean, sailing into history as the first vessel to complete a successful circuit of the globe.

According to Jonathan Raban, it was the invention of the magnetic compass, over a thousand years ago, that gave mankind the hubris to imagine that it was possible to sail in straight lines across the sea. 'Once upon a time,' he wrote, 'people made their way across the sea by reading the surface, shape, and colours of the water', but once the compass became established on the quarterdeck, everything changed:

Sebastian Munster's map of the Americas, 1561, featuring Magellan's name for the Pacific, *Mare pacificum*, and the Strait of Magellan, labelled *Frenum Magaliani*.

the whole focus of the helmsman shifted, from the sea itself to an instrument eighteen inches or so under his nose. Suddenly he no longer needed to intuit the meaning of the waves; he had become a functionary, whose job was to keep the ship at an unvarying angle to the magnetized pointer with its scrolled *N*.[16]

Portolan chart, probably drawn in Genoa, mid-14th century. The chart covers the Mediterranean Sea from the Balearic Islands to the Levantine coast. It is the oldest cartographic artefact in the U.S. Library of Congress.

Medieval portolan charts showed direct sailing routes in the form of rhumb lines drawn with rulers across flat seas of vellum, but navigators used to sailing by tides and currents would have known how meandering such courses would really look if drawn accurately in two dimensions. Portolan charts, named after early Italian pilot books, or *portolano*, appeared soon after the invention of the magnetic compass in the early twelfth century. They primarily represented areas of the Mediterranean and Black Sea, with some covering the Atlantic as far as Ireland and the west coast of Africa, and demand for them grew as trade by sea

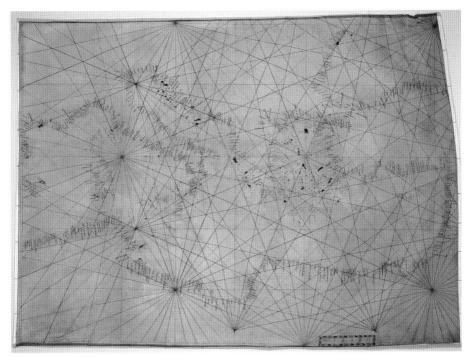

increased over the following centuries. European trading ports such as Genoa, Venice, Majorca and Barcelona cooperated in sharing maritime and coastal information gleaned from returning seafarers; from repeated revisions and new compass surveys, the reliability of the portolan charts soon surpassed all earlier maps, in spite of the fact that they were based on projections that assumed that degrees of longitude were equal to degrees of latitude. This assumption made little difference in a Mediterranean or equatorial context, but it caused serious distortions in maps that ventured further north or south. At the equator, one degree of longitude measures around 111 km (69 mi.), roughly equivalent to the 110 km of one degree of latitude, but at higher latitudes, say, at 80 degrees north, in the Greenland Sea, the difference is dramatic, with one degree of longitude shrinking to around 19 km (12 mi.) against the 111 km of one degree of latitude. Ships became routinely lost as a result of the flattening effect of a globe depicted on flat ruled charts. Early Pacific navigators, sailing without compasses, consistently demonstrated a greater understanding of oceanic pathways than later compass-and-quadrant-led seafarers, who were deprived by their own technology of an ancient intimacy with the sea. Worse, as Greenville Collins's *Coasting Pilot* (1693) pointed out, such instruments grew 'dull' with use at sea, the metal pin on which a compass needle rests being prone to wearing blunt, 'and the Compass being not quick in traversing'; he advised sharpening it 'with a Hone or fine Whet-stone with care and exactness' at the start of every voyage, though it's not clear how many ships' commanders ever heeded his advice.[17]

The drawbacks of trusting to a compass-driven course were powerfully illustrated in Joseph Conrad's seafaring novella *Typhoon* (1902), in which the stubborn captain, Thomas Mac-Whirr, is unwilling to deviate one degree from his predetermined line, and drives the ss *Nan-Shan* directly into a cyclone in the South China Sea. As the storm builds, the first mate, Jukes, urges MacWhirr to turn the steamer eastwards, to meet the perilous cross-swells running in advance of the typhoon:

'Head to the eastward?' he said, struggling to sit up. 'That's more than four points off her course.'

'Yes, sir. Fifty degrees. . . . Would just bring her head far enough round to meet this. . . .'

Captain MacWhirr was now sitting up. He had not dropped the book, and he had not lost his place.

'To the eastward?' he repeated, with dawning astonishment. 'To the . . . Where do you think we are bound to? You want me to haul a full-powered steamship four points off her course! . . . If I didn't know you, Jukes, I would think you were in liquor. Steer four points off. . . . And what afterwards? Steer four points over the other way, I suppose, to make the course good. What put it into your head that I would start to tack a steamer as if she were a sailing-ship?'[18]

The epitome of a modern technological mariner divorced from nature, Conrad's captain views his ship as a commercial object, designed to cross the ocean in inflexible straight lines. As Jonathan Raban observes, 'the whirling cyclone that Conrad brews up to engulf the stupid captain and his crew is the ocean's revenge for the hubris of the steam turbine and the ruled line on the chart.'[19] For Raban, the only true way to understand the sea was to read its movements directly, to look to the swells and currents and navigate accordingly, like the Santa Cruz islanders of Melanesia, who recognized and named an array of Pacific swell patterns: the *hoahualoa*, the 'Long Swell' from the southeast trade winds; the *hoahuadelahu*, produced by the northwest monsoon; and *hoahuadelatai*, the 'Sea Swell', a distant echo of the northeasterly trade winds.[20] This was lifesaving knowledge based on close observation of the sea, codified in a highly expressive maritime language, the spread and development of which has been the subject of a wealth of scholarly and writerly attention.

Shipwrecks

On 2 June 1609 a seven-ship supply fleet set sail from Plymouth, Devon, destined for the colonial settlement of Jamestown,

The coat of arms of Bermuda features the wrecking of the French ship *Edward Bonaventura* in 1591. The motto, *Quo fata ferunt* (Whither the fates carry us), is derived from Book v of Virgil's *Aeneid*.

Virginia. Six weeks later, the fleet encountered a severe storm (possibly a hurricane) off Bermuda, and the ships became separated. After three days of battering, the 300-ton flagship *Sea Venture* – Britain's first purpose-built emigrant ship – began taking on water, at which the commander, Admiral Sir George Somers, had the ship deliberately driven onto the reefs on what is now St George's Island. All 150 passengers and crew made it safely onto the island, but the ship itself was damaged beyond repair.

Over the following nine months the castaways built two new pinnaces, *Deliverance* and *Patience*, from local Bermudan cedar, along with ironware and rigging salvaged from the wreck. By early May 1610, the two beach-built vessels were ready to set sail, and two weeks later *Deliverance* and *Patience* arrived at Jamestown with all the survivors on board, only to find a handful of colonists dying of illness and starvation. The whole fateful episode was recorded by the writer and adventurer

William Strachey, whose eyewitness account, the *True Reportory of the Wracke, and Redemption of Sir Thomas Gates Knight* (1625), had circulated in manuscript among members of the Virginia Company back in London, including the Earl of Southampton, a friend of William Shakespeare. Today, most Shakespeare scholars agree that the *True Reportory* must have come into the playwright's hands in the autumn of 1610, when he was beginning work on his final play, *The Tempest* (1611), for there are enough similarities between the two texts to demonstrate that Shakespeare was familiar with Strachey's account of the 'most dreadfull Tempest' that had scuppered the *Sea Venture*. For example, Strachey's description of the 'Tortoyse' (probably a sea turtle) that the shipwrecked emigrants ate on Bermuda as 'such a kind of meat, as a man can neither absolutely call Fish nor Flesh' is echoed by Prospero calling Caliban 'thou tortoise', at which Trinculo asks if he is 'a man or a fish?', along with numerous other echoing phrases, such as Strachey's 'sharpe windes blowing Northerly' versus Prospero's 'sharp winds of the north', or Strachey's long description of St Elmo's fire (an electrical weather phenomenon) 'striking amazement' along the main mast, 'streaming along with a sparkling blaze . . . shooting sometimes from Shroud to Shroud' versus Ariel's mischievous account of the same occurrence:

Now in the waist, the deck, in every cabin
I flamed amazement. Sometimes I'd divide
And burn in many places – on the topmast,
The yards and bowsprit would I flame distinctly,
Then meet and rejoin.[21]

For Shakespeare, the shipwreck had proved a tried and tested theatrical device through which characters could be exiled to faraway locations in which they might then reinvent themselves, as deployed in earlier plays such as *The Comedy of Errors* (*c*. 1594) and *Twelfth Night* (1601–2). In both, the shipwreck acts as the catastrophic force that separates the protagonists from their families – and because the protagonists of both these

Overleaf: Romantic-era *Sturm und Drang* exemplified in Knud Andreassen Baade's *The Wreck*, 1830s, oil painting.

earlier plays are twins, they are also separated from aspects of themselves, as hinted at in an early speech of Antipholus of Syracuse (from one of the two sets of separated twins whose fortunes propel *The Comedy of Errors*):

I to the world am like a drop of water
That in the ocean seeks another drop,
Who, falling there to find his fellow forth,
Unseen, inquisitive, confounds himself. (1.2)

The two Antipholuses eventually meet and rejoin in the final scene, like the two drops of seawater introduced in the first. In common with the majority of early shipwreck narratives, these plays characterized shipwrecks as instances of endurance and hope rather than as purely disastrous events. A shipwreck, after all, was a common occurrence, and given that most ships went to sea again and again, the likelihood of eventual wrecking was high. But shipwreck narratives such as the *True Reportory* tended to show a crisis at sea overturned by the eventual restoration of order – the ship's rebuilding and the crew's perseverance – even though, in this case, the voyage itself ended in failure.

Shakespeare's storm-driven dramas had many other antecedents, including medieval productions of the Noah story that were staged by maritime guilds, sometimes with striking verisimilitude. In the late fourteenth-century York shipwrights' pageant *The Building of the Ark*, for example, God instructs Noah to select the tallest trees, cut their ends square and use board-and-batten construction with nails and glue, effecting Noah's transformation through grace into a skilled medieval shipwright. The details of the clinker-building process, with wooden keel and cross-ribs overlaid with 'strakes' (rows of horizontal overlapping planks to make up the hull), were so accurate that productions of the play featured actual on-stage vessels taking shape before the audience's eyes.[22] The Wakefield *Noah and the Ark* followed an equally detailed course in do-it-yourself shipcraft, with Noah sounding the water's depth while his wife takes the tiller:

A view from the rigging on board the Scottish-built three-masted barque *Garthsnaid*, c. 1920. Crew members can be seen securing a section of the foresail that had come free from the gaskets in heavy weather.

NOAH: Wife, tent the stere-tree,
And I shall asay
The deepnes of the see
That we bere, if I may.
NOAH'S WIFE: That shall I do ful wisely.
Now go thy way,
For apon this flood have we
Flett many day.[23]

Such plays served to familiarize their audiences with maritime subjects, as well as with nautical terminology, some of it surprisingly technical. As the following section shows, the creation and dissemination of such languages of the sea flourished over the centuries, involving a range of discourses, spoken and written, from the purely functional to the highly literary, which together formed a distinctive, salt-tinged vernacular that by the mid-nineteenth century was known as 'Jack-speak'.

Language of the sea

For much of human history the sea has been what social anthropologist Tim Ingold termed a 'taskscape', a space delimited by
the processes of work and endurance.[24] Like all workplaces the
sea has generated an array of languages and argots associated
with the knowledges, experiences and rituals of maritime and
naval life, as well as with establishing communities of trust and
belonging. As the anthropologist John Mack observed, 'language – rather than simply sailing skills – comprises one of the
first elements of initiation into seafaring', and he cited the
example of Jack Cremer, an early eighteenth-century American
sailor, who dated his acceptance into the ship-board community
to the moment when his shipmates 'began to learn me to call
names, which was the first Rhudiment of that University'.[25] To
learn the working language of ship-board life was to join a linguistic fraternity that mixed technical vocabulary with
specialized slang to create an alluringly arcane vernacular. Much
of that technical vocabulary had been derived from the Dutch
language, which spread across the world throughout the sixteenth and seventeenth centuries in the wake of the dominance
of the United Provinces as a global seafaring nation. 'All hands
on deck', for example, was borrowed directly from '*alle hens aan
dek*', while dozens of familiar nautical terms in English, including 'buoy', 'bulwark', 'commodore', 'yacht', 'yawl', 'keel-haul',
'sloop', 'hoist', 'schooner', 'bowline', 'cruise' and 'iceberg', all began
their spoken life in Dutch. But when Dutch maritime suprem-
acy began to decline during the early eighteenth century, and the
British proceeded to rule the waves instead, English soon established itself as the new (though still heavily Dutch-inflected)
international argot of the sea.

For the authors of the many maritime dictionaries that
have appeared over the centuries, beginning with Sir Henry
Mainwaring's *Seaman's Dictionary; or, Nomenclator Navalis*
(1644), their purpose seemed as much celebratory as practical.
Admiral William Henry Smyth, for example, prefaced his
seven-hundred-page *Sailor's Word-Book* (1867) with a fond

endorsement of the picturesque precision of so-called Jack-speak:

> The predilection for sea idiom is assuredly proper in a maritime people, especially as many of the phrases are at once graphic, terse, and perspicuous. How could the whereabouts of an aching tooth be better pointed out to the operative dentist than Jack's ''Tis the aftermost grinder aloft, on the starboard quarter.'[26]

As so many of Smyth's examples showed, ship-board language was both memorable and precise, 'a flawless thing for its purpose', as Joseph Conrad described it.[27] Ambiguity of expression might be tolerable on land, but it could be fatal at sea, where clear and unmistakable orders needed to be given to groups of active people, crowded onto a deck, often in challenging weather conditions. But such jargon can only be learned on the job, and for any landlubber who hitches a ride on a professional vessel, language (rather than seasickness) is often the greatest obstacle. For Geoff Dyer, whose two-week residence on an American aircraft carrier in the Persian Gulf was outlined in *Another Great Day at Sea* (2015), his failures to understand nautical terminology were, he concluded, 'failures at the level of the noun, exceeded by systematic failures at the level of the verb: what these nouns – these various parts – *did*', while in Redmond O'Hanlon's *Trawler* (2003), his account of joining a midwinter voyage to the Icelandic fishing grounds, ship-board vocabulary was a similar source of bemusement:

> 'What did you mean – *lumps?*'
> 'Lumps? Waves! To a trawlerman a big wave is never a wave, it's a lump. Cuts it down to size, I suppose. *Wave* is too serious. You don't want the sea to know you're frightened, do you?'[28]

Whalers once employed a similarly belittling strategy by referring to whales as 'fish' (as Melville does throughout *Moby-Dick*), just as lumberjacks liked to refer to the tallest trees as 'sticks'. As the 'lumps'/'waves' substitution shows, seafaring language is shaped by propitiatory as well as by practical considerations, with alternative vocabularies arising in the service of long-established threshold taboos; the word 'langlugs', for example, remains a widely used Anglophone substitute for 'hare', a particularly ill-omened animal on board ship, along with rabbit, cat, dog and fox. So great was a seafarer's animus against naming a four-legged animal at sea that most coastal communities developed repertoires of substitute, non-taboo terms. In Shetland, philologists have identified no fewer than eighteen propitiatory words for 'horse', thirteen for 'pig', eleven for 'sheep' and seven for 'cow', all for use at sea as so-called 'luckywords', while Faroese developed as many as 22 words for 'cat'.[29] In fact Faroese has been found to comprise two distinct, parallel languages, a sea language and a land language (known as the 'kilnhouse' version), with each spoken only in context, though often by the same people, who might well be fishers at sea and farmers on land. The people of Fair Isle similarly employed two sets of place names, one sea based and the other land based, with 'superstition guaranteeing their separation, those place-names used from the sea never being used from the land'.[30]

As Redmond O'Hanlon discovered, modern fishing boats still run on such ancient codes of superstition and belief, and his first conversation before setting foot on the trawler concerned what he could and could not say and do during his time on board: he must not wear green, he must turn back if he sees a minister of religion on the way to the ship; he cannot leave harbour on a Friday: 'Your wife, for instance, whatever happens, she must *never* use the washing machine on the weekend before you go. Because it's like the sea, the whirlpool – she'll be washing your soul away.'[31]

Breaking any of these taboos would spell misfortune for the ship, although the age-old remedy of touching cold iron was still in place, a counteraction shared by fishing cultures across the North Atlantic, with Scandinavians touching *kallt järn*, and the Shetlanders *cauld ærn* in cases of linguistic transgression (and there is always plenty of cold iron at hand on a trawler). Inevitably, one of the final passages in his book sees O'Hanlon unwittingly jinxing the boat by mentioning rabbits, before calling out 'cold iron' and grabbing the rail.

The consequences of breaking a language taboo could be serious. The legend of the *Flying Dutchman* had its origins in the blasphemy of its captain, Vanderdecken, during a storm at sea, the captain being then 'condemned by the Almighty to sail the seas continually'.[32] His vessel, the *Flying Dutchman*, unable to make port, becomes over time a ghost ship with a spectral crew, sightings of which inspire terror among the crews of other, non-phantasmal ships. Frederick Marryat told a compelling version of the legend in his novel *The Phantom Ship* (1839), in which a close encounter with the cursed apparition presages disaster for those sailors ill-fated to run into her:

> Not a breath of wind was on the water – the sea was like a mirror – more and more distinct did the vessel appear, till her hull, masts, and yards were clearly visible. They looked and rubbed their eyes to help their vision, for scarcely could they believe that which they did see. In the centre of the pale light, which extended about fifteen degrees above the horizon, there was indeed a large ship about three miles distant; but, although it was a perfect calm, she was to all appearance buffeting in a violent gale, plunging and lifting over a surface that was smooth as glass, now careening to her bearing, then recovering herself . . . She made little way through the water, but apparently neared them fast, driven down by the force of the gale. Each minute she was plainer to the view. At last, she was seen to wear, and in so doing, before she was brought to the wind on the other tack, she was so close to them that they could distinguish the men on board: they could see the foaming water as it was hurled from her bows; hear the shrill whistle of the boatswain's pipes, the creaking of the ship's timbers, and the complaining of her masts; and then the gloom gradually rose, and in a few seconds she had totally disappeared![33]

The legend was mirrored in reality by the unsettling discovery, in December 1872, of an American merchant ship drifting under sail in the North Atlantic without a soul on board. The ship was still seaworthy and well provisioned, showing no signs of struggle or misfortune; it had apparently been abandoned mid-ocean by its captain and crew, none of whom was ever seen again. Rumours about the ship abounded; had she been somehow cursed? Her first captain had died midway through her maiden voyage in 1861, at the end of which her replacement captain accidentally rammed and sank an English brig in the Strait of Dover. The fated vessel had originally been named the *Amazon*, but in 1868 she had been sold, her new owners renaming her the *Mary Celeste*. According to maritime superstition, special care was needed when it came to the business of renaming a ship, since the given name of every vessel was believed to be recorded by the 'Ledger of the Deep', a kind of bureaucratic manifestation of Neptune. To fail to notify the Ledger of a change of name was likely to invoke his wrath, so all traces of a ship's previous identity needed to be removed, not just from the vessel herself, but from all associated documents, badges and receipts. A solemn ceremony would then be needed to safely rechristen the ship.

Had some kind of naming code been breached in the case of the *Mary Celeste*? After the discovery of the abandoned ship made headlines around the world, shipyard workers recalled how the *Amazon* (as she was then) had got ominously stuck on the slipway during her maiden launch, needing a dozen men to haul her into the water. 'The craft seemed possessed of the devil to begin with, but where she got it I don't know,' as the son of the ship's original builder, Joshua Dewis, was quoted as saying.[34] Following her rescue, the restored *Mary Celeste* continued in service under a sequence of notably unlucky owners, all of whom lost money by her, and in 1885 she was deliberately wrecked off the coast of Haiti as part of an attempted insurance fraud. Her last captain, Gilman Parker, was arrested and charged with fraud as well as 'the wilful casting away of a ship', but the trial collapsed, and Parker walked free, dying in a poorhouse three months later. The cursed *Mary Celeste* has remained an icon of fear and

The *Nautilus* sails through a shipwreck; engraved illustration from an early edition of Jules Verne's *Twenty Thousand Leagues Under the Seas*.

fascination among seafarers ever since, and a ready-made precedent for any abandoned ship found at sea, such as the 80-metre (260 ft) tanker *Jian Seng*, found adrift in the sea off Queensland, Australia, in 2006, with all identifying marks having apparently been erased by the vanished crew. The coastguard's attempts to discover the mystery vessel's origins failed, and the ghost ship was eventually scuttled in deep water, a contemporary *Mary Celeste*.

Although the figure of the seaman had long been a cultural fixture – the drunken sailor was already a stereotype in Chaucer's time – it was not until the Napoleonic era that the richness of nautical life and language found literary representation. John Davis's *The Post-Captain; or, the Wooden Walls Well Manned* (1805) is considered 'the parent of all our nautical novels', and was filled with evocative maritime speak, such as the captain's rhapsody to the wind of fortune: 'Blow my good breeze! Fill my sails! driver and ring-tail, spritsail and sprit-topsail! Royals and sky-scrapers! flying jib and jib of jibs!', or midshipman Echo's echoing instructions along the chain of command:

Boatswain's mate! boatswain's mate! I say, you boatswain's mate! send the after-guard aft here to the main-topsail-haliards. Corporal of marines! send the marines aft on the quarter-deck, to clap on the main-topsail-haliards. Master at arms! go down below and send all the idlers up! Send all the idlers up! Do you hear, there, master at arms? Send all the idlers up! . . . After-guard! I don't see the after-guard coming aft! Where's the captain of the after-guard? Pass the word there in the waist for the captain of the after-guard![35]

According to Smyth's *Sailor's Word-Book*, 'idlers' referred to members of the crew not required on night-watch, while the 'waist' was the central deck of a ship, where non-combatants mostly worked. 'Idlers' and 'wa[i]sters' both entered the general language as derogatory terms, alongside the likes of 'loose cannon' and 'landlubber', such words having become culturally familiar over the course of the lengthy naval campaigns of the Napoleonic wars.

Maritime language would also become associated with a strong storytelling tradition, with the trope of the yarning sailor a key feature of nautical lore. In Dickens and Collins's Christmas tale 'A Message from the Sea' (1860), the resurrected castaway Hugh Raybrock is distinguished from his fellow seafarers by a remarkably un-tarlike inability to tell his story: 'A sailor without a story! Who ever heard of such a thing?', as one of his hosts exclaims, underlining the widely held view of storytelling as one of the seafarer's chief characteristics.[36] For Walter Benjamin, writing in the 1930s, 'the trading seaman' was an archetypal master storyteller, who returned from overseas with tales of adventure that also communicated 'the lore of faraway places: "When someone goes on a trip, he has something to tell about," goes the German saying, and people imagine the storyteller as someone who has come from afar.'[37]

One of the most compelling examples of the maritime storyteller was the loquacious Captain Kearney, from Frederick Marryat's fourth (and best) novel, *Peter Simple* (1834). Kearney is a notorious teller of incredible stories, 'the greatest liar that ever walked a deck', who only ever told the truth by accident; even as he lay dying in a naval hospital in Nova Scotia, he insisted upon dictating an entirely fictitious will, leaving his imaginary estate '(let me see what is the name of it?) Walcot Abbey, my three farms in the vale of Aylesbury, and the marsh lands in Norfolk' to his equally imaginary children, William Mohamad Potemkin Kearney and Caroline Anastasia Kearney, as well as to his fictitious wife, Augusta Charlotte Kearney ('she was named after the Queen and Princess Augusta, who held her at the baptismal font').[38] Beneath the comedy there's a sharp poignancy to this penniless, unmarried sailor fantasizing on his deathbed about his landed family possessions, most of which lay nowhere near the sea.

Kearney is one of a host of vividly drawn characters who populated Captain Marryat's twenty naval novels, beginning with *The Naval Officer* (1829), and ending, more than 3 million words later, with *The Privateersman* (1846). The books were based on Marryat's own long career at sea; he had joined the

Navy, aged fourteen, a year after the Battle of Trafalgar, serving as a volunteer midshipman on the frigate *Impérieuse*. Marryat's first week at sea nearly proved his last: in November 1806 a storm drove the *Impérieuse* on to rocks off the French island of Ushant. Marryat never forgot 'the grating of the keel as she was forced in; the violence of the shocks which convulsed the frame of the vessel ... and then the enormous waves which again bore her up and carried her clean over the reef', and more than twenty years later he drew on the incident for the celebrated storm sequence in *Peter Simple*:[39]

> A tremendous sea, which appeared to have risen up almost by magic, rolled in upon us, setting the vessel on a dead lee shore. As the night closed in, it blew a dreadful gale, and the ship was nearly buried with the press of canvas which she was obliged to carry: for had we sea-room, we should have been lying-to under storm stay-sails; but we were forced to carry on at all risks, that we might claw off the shore. The sea broke over us as we lay in the trough, deluging us with water from the forecastle aft, to the binnacles; and very often as the ship descended with a plunge, it was with such force that I really thought she would divide in half with the violence of the shock ... It really was a very awful sight. When the ship was in the trough of the sea, you could distinguish nothing but a waste of tumultuous water; but when she was borne up on the summit of the enormous waves, you then looked down, as it were, on a low, sandy coast, close to you, and covered with foam and breakers.[40]

As Joseph Conrad observed in his essay 'Tales of the Sea' (1898), Marryat may not have loved the sea, but the sea loved him without reserve: 'to this writer of the sea the sea was not an element: it was a stage [which] gave him his professional distinction and his author's fame', although as Conrad also observed, Marryat's novels were at their best when the action was on the water, the stories floundering like amphibious creatures whenever they touched dry land.[41]

Frontispiece to George Cruikshank's naval satire *The Progress of a Midshipman, Exemplified in the Career of Master Blockhead* (1820), a series of coloured etchings based on original drawings by Captain Frederick Marryat. This image introduces the nervous young hero as he crosses the sea towards the Temple of Fame.

The kinds of picturesque Jack-speak invoked in the novels of Marryat and Conrad (and in those of their followers, notably C. S. Forester and Patrick O'Brian) have now all but disappeared, replaced by an abbreviated form of maritime English known as 'SeaSpeak': an artificial sublanguage that was formally adopted by the International Maritime Organization (IMO) in 1988, and revised in the 1990s to accommodate changes in communication technology. So today, a typical conversation between passing ships anywhere in the oceans might go something like this:

Nippon Maru: Gulf Trader, Gulf Trader. This is Nippon Maru, Juliet-Sierra-Alpha-Alpha. Nippon Maru, Juliet-Sierra-Alpha-Alpha. On VHF channel one-six. Over.
Gulf Trader: Nippon Maru, Juliet-Sierra-Alpha-Alpha. This is Gulf Trader. Alpha-Six-Zulu-Zulu. Over.
Nippon Maru: Gulf Trader. This is Nippon Maru. Switch to VHF channel zero-six. Over.

Gulf Trader: Nippon Maru. This is Gulf Trader. Agree VHF channel two-six. Over.

Nippon Maru: Gulf Trader. This is Nippon Maru. Mistake. Switch to VHF channel zero-six. I say again. Switch to VHF Channel zero-six. Over.

Gulf Trader: Nippon Maru. This is Gulf Trader. Correction. Agree VHF channel Zero-six. Over.[42]

The restricted vocabulary and standardized phrasing of SeaSpeak ('mistake'; 'correction'; 'say again') was designed to facilitate communication across the medium of very high frequency (VHF) radio, where clarity is often blurred by poor reception at sea, as well as across multiple language barriers. SeaSpeak was a linguistic response to the rise of containerization, the workforce for which is recruited from almost every nation on Earth, as hinted at in the title of the language's official handbook, *SeaSpeak Training Manual: Essential English for International Maritime Use* (1988).[43]

Containerization has transformed not only the language of the sea, but the entire character of seaborne transportation.

A waterspout at sea, from Camille Flammarion's *L'atmosphere* (1873).

The view out to sea from the Coast Watch Station on the roof of Martello Tower 'P', Felixstowe, Suffolk.

There are currently more than 50,000 container ships afloat, 'the most independent objects on earth', as William Langewiesche describes them in *The Outlaw Sea* (2005), his study of the largely unregulated world that these vessels inhabit as they criss-cross the oceans under flags of convenience, with more than 90 per cent of the world's traded goods laden on their decks.[44] The flag of convenience system began in the 1920s, when American shipowners, frustrated by increasing domestic regulation and rising wage costs, began to register their ships in tax-free Panama instead. The practice spread, and Panama is now the largest maritime nation on Earth, followed by Liberia and the tiny Marshall Islands, although no coastline is actually required to register as a shipping nation. There are ships that hail from La Paz, in landlocked Bolivia; there are ships that hail from the Mongolian desert.[45] In the opening pages of *Typhoon* (1902), Joseph Conrad offered a prescient commentary on the practice, with the first mate reacting superstitiously to the new Siamese

(Thai) flag under which Captain MacWhirr agrees to command the Scottish-built steamer, the *Nan-Shan*: 'The first morning the new flag floated over the stern of the *Nan-Shan* Jukes stood looking at it bitterly from the bridge. He struggled with his feelings for a while, and then remarked, "Queer flag for a man to sail under, sir,"' before predicting, correctly, that no good would come to the vessel.[46]

One of the themes of Conrad's story had been the diminishing romance of seagoing, embodied by MacWhirr's inelegant cargo steamer, which, unlike a ship under sail, had been built to steer a straight compass course from one port to another, foreshadowing the coming age of mass containerization. It mirrored a broader naval decline since the end of the Napoleonic era, when command of the seas had required a well-manned fleet, rather

Dawn breaks over the Port of Rotterdam, *c.* 1915. A century later, Rotterdam has become the largest seaport in Europe, covering an area of 105 square km (41 sq. mi.), and handling nearly 500 million tonnes of cargo every year.

Containerization: more than 40 per cent of UK imports come via the Port of Felixstowe.

than a mainly strategic maritime presence. As commercial cargoes expanded, purpose-built container ports such as Fremantle Harbour in Western Australia, or the Port of Felixstowe in Suffolk, England, grew to handle enormous quantities of containerized cargo in largely automated docking systems, although in recent decades the maritime world's centre of gravity has shifted to Asia, where vast deep-water container ports, such as the 3,600 square km (1,400 sq. mi.) Port of Shanghai, on the east China coast, dwarf those of the European maritime nations. In January 2015 the world's then largest container ship, the Chinese-owned MV *CSCL Globe*, berthed at Felixstowe with a record load of 19,000 6-metre (20 ft) shipping containers, carrying as many items as there are people in the British Isles: arriving with, literally, something for everyone.

A 2010 film essay, *The Forgotten Space* (dir. Noël Burch and Allan Sekula), explored this transformation of the commercial seas, featuring interviews with those whose working lives are shaped by the now-ubiquitous steel shipping container – which first appeared in the United States in the 1950s – and asking whether containerization has 'turned the sea of exploit and adventure into a lake of invisible drudgery'. The intermodal container, which can be transferred easily between different forms of transportation, from ship to train to HGV, transformed the sea into a global trading floor, anonymizing the cargo: ship workers no longer know what they are carrying, with the containers locked for the duration of the voyage, except for refrigerated containers, the temperatures of which are meant to be checked daily, but are often overlooked in the frenzied conditions of onboard life.

Despite the automatization of much of the work, the sea remains a dangerous working environment, especially in waters as crowded as the English Channel or the Suez Canal. As Rose George observes in *Deep Sea and Foreign Going* (2013), a study of life on board a modern container ship, 'every minute without incident at sea is simply a minute in which danger has been avoided', whether in the form of other vessels and unseen obstacles, including discarded shipping containers, or from bad

The view from the deck of an ocean-going container ship, in a still from the haunting essay film *The Forgotten Space* (dir. Noël Burch and Allan Sekula, 2010).

In March 1978 the supertanker *Amoco Cadiz* ran aground off the coast of Brittany during a gale, releasing nearly 220,000 tonnes of crude oil into the sea: it remains one of the worst oil spills on record.

weather and the unpredictable caprices of the sea itself.[47] In March 1978, for example, the supertanker *Amoco Cadiz* encountered gale-force winds and unusually high seas during a storm in the English Channel. After a particularly heavy wave damaged the ship's rudder, rendering the vessel unmanoeuvrable, the tanker was blown by Force 10 winds onto the rocky coast of Brittany, where it broke apart, releasing nearly 220,000 tonnes of crude oil into the sea, leading to a devastating loss of marine life. Fatal collisions at sea are also surprisingly common: in June 2017 uss *Fitzgerald*, a naval destroyer, was accidentally rammed by a 29,000-tonne container ship off the coast of Japan, killing seven members of the destroyer's crew; two months later, another American warship, uss *John S. McCain*, was rammed by an oil tanker off the coast of Singapore, with the loss of a further ten lives. The official report into the second collision blamed the

accident and loss of life on 'insufficient training' of the warship's crew alongside inadequate bridge operating procedures.

But if low pay and unsafe working conditions are the lot of the commercial seafarer today, things were not much better back in the age of sail. Among the many psychological conditions that afflicted European mariners was 'calenture', a powerful hallucination, caused by heatstroke, in which a sailor (usually while alone on deck) became convinced that the surrounding swells were the green fields of home, and so jumped overboard to his death. In 1829 a French naval vessel, the *Dunesque*, en route to Rio de Janeiro, reportedly lost one hundred of its six-hundred crew to the syndrome, and it has recently been argued that the naval term 'missing at sea', which historically accounts for around 4 per cent of deaths in the service, can largely be attributed to calenture, 'which is not to be thought of as suicide, since death was not the object, but the hallucinations were such that the sailor felt "powerless to resist" and "hypnotically attracted" into leaping off the ship and into the sea'.[48] The condition usually struck in calm tropical waters

William Bradford, *Shipwreck off Nantucket*, c. 1860–61, oil on canvas. Bradford had been an eyewitness to the wreck of the whale ship *Nantucket*, at the entrance to Vineyard Sound in August 1859. The dramatic effect of the heavy sea and tilting ship shows the influence of Dutch marine art on Bradford's style.

under cloudless skies, after a vessel had been in open sea for at least a week with no visible break on the horizon.

Incidents of calenture have not been reported since the nineteenth century, but the condition remains notorious, partly from the striking descriptions to be found in *Moby-Dick* (Chapter 114, 'The Gilder') and in Poe's *Narrative of Arthur Gordon Pym of Nantucket* (1838), in which the sea appears as 'waving meadows of ripe grain', and partly from Jonathan Swift's famous invocation of the condition in his satirical poem, 'Upon the South Sea Project', of 1721:

> So, by a calenture misled,
>   The mariner with rapture sees
> On the smooth ocean's azure bed
>   Enamelled fields and verdant trees;
>
> With eager haste he longs to rove
>   In that fantastic scene, and thinks
> It must be some enchanted grove,
>   And in he leaps, and *down* he sinks.[49]

5 The Sea in Art and Music

If there is poetry in my book about the sea, it is not because
I deliberately put it there, but because no one could write truthfully
about the sea and leave out the poetry.
Rachel Carson, 1952

In the summer of 1972, the conceptual artist Susan Hiller
bought a vintage postcard in a British seaside town bearing the
caption 'Rough Sea, Weston-Super-Mare'. 'There was a picture
of an extraordinary wave and there were words,' Hiller recalled.
'I looked at it for a long time.'[1] A few weeks later, Hiller visited
Brighton, in Sussex, where she found another postcard of a
wave breaking over the local promenade, with a similar 'Rough
Sea' caption. If there were two of them, she thought, there must
be more, and she set about searching for other examples of
these crashing-wave images, 'a genre of postcard that had all
sorts of fascinating aspects. I thought of the cards as miniature
artworks.'

Four years on, Hiller had collected more than three hundred
'Rough Sea' postcards from around the British Isles, most of
them dating from the early 1900s, and had organized them into
a curated artwork, entitled *Dedicated to the Unknown Artists*, in
tribute to the anonymous photographers, painters and hand-
tinters who had created the original artefacts. The collection,
which was first exhibited at the Gardner Centre Gallery,
Brighton, in 1976 (and is now part of the Tate collection) was
arranged in grid formation over fourteen large panels, alongside
an annotated map of Britain marking the coastal locations
featured on the cards. A number of the senders' messages were
also reproduced. One, posted from Clacton-on-Sea, reads, 'it
has been really like this today, had splendid time,' while another,
picturing a rough sea at the Giant's Causeway, Antrim, states:

ROUGH SEA. BRIGHTON.

An early 20th-century 'Rough Sea' postcard. These were popular as souvenirs sent from windswept British resorts.

'it has been wet ever since I came.'[2] Hiller has spoken of her fascination with these competing representational languages – verbal and visual – and how they give rise to 'a series of paradoxes involving the unexpressed but intended vs. the expressed but unintended'.[3] The cards' repeated seaside motifs, with their connotations of leisure and romance amid the incursion of violent natural forces, contrasts with the strict conceptual methodology of the minimalist grid formation, emphasizing the implied contradiction between the implacable power of the sea and the dull respectability of out-of-season resorts, in which vigorous plumes of water cascade across tidy rows of guest houses. Virginia Woolf memorably described seeing just such a sight at Seaford, Sussex, in October 1937, 'the sea over the front: great spray fountain bursting to my joy over the parade and the lighthouse. Right over the car . . . a great curled volume roughened of water. Why does one like the frantic the unmastered?', a question that is answered with potent visual eloquence by Susan Hiller's work.[4]

Hiller's ever-growing postcard collection gave rise to two further series: *Rough Seas* (1982–2019), a sequence of tinted prints that explored the relationship between landscape painting and photography, and *On the Edge* (2015), an expanded version of *Dedicated to the Unknown Artists*, featuring 482 different postcard views of 219 seaside locations, its topographical arrangement offering a conceptual circumnavigation of the weather-beaten coastline of Britain.

Hiller's life-long fascination with storm waves speaks to a recurring creative and emotional response to the sea in motion. In a long chapter on 'The Truth of Water' in volume 1 of *Modern Painters* (1843), the art critic John Ruskin argued that the sea poses a particular challenge to visual representation, its restless turbulence being too often conveyed as mere formless disorder. 'The sea *must* be legitimately drawn,' he wrote, 'it cannot be given as utterly disorganised and confused,' and he urged artists to pay close attention to what he defined as the sea's essential 'fury and formalism'.[5] For Ruskin, as for Hiller, the sea in motion was chaotic yet ultimately readable, being alive with a dynamic, self-renewing power that put static visual art forms such as painting and still photography to the test. Ruskin urged artists to rise to the challenge by developing a visual understanding of water on the move, and learning the difference between wind action on an estuary and its effects out at sea, between the breaking of rollers on a gently shelving beach and the pounding of wind-driven waves on rock; although, when it came to depicting a breaking wave, he conceded, 'it is that last crash which is the great taskmaster. Nobody can do anything with it.'[6]

Ruskin's views were seconded by the Victorian geologist David Ansted, in an influential essay on 'The Representation of Water' that was published in the *Art Journal* in 1863; as had Ruskin, Ansted observed that waves were a problem for visual artists, being characterized by 'variety without end', but that close attention to their movement would be well rewarded:

> From the gentlest ripple to the most violent disturbance the gradations are infinite; and the waves vary, not only with their magnitude, but with the depth of the water in which

they are formed. Every wave surface, besides having its own height and width as a wave, is also covered with small ripples, so that perfectly smooth surfaces of water are rare and exceptional appearances [but] they afford studies of a high order, teaching some most difficult and little known truths.[7]

Taking both Ruskin's and Ansted's cue, this chapter will explore some of the many ways that artists, film-makers, composers and musicians have responded, whether visually or conceptually, to the sea's timeless fury and formalism, as well as to its networks of meanings and motifs, to its difficult and little-known truths.

The sea in visual art

One of the earliest depictions of the sea in European art can be found in the Bayeux Tapestry, a 70-metre-long (230 ft) embroidered cloth created in England around 1077. While ships are shown in some detail in the scenes depicting King Harold's shipwreck off the coast of Normandy, and William's subsequent invasion fleet, the sea itself is depicted in simple schematic form, by a series of undulating lines arranged in parallel rows; as the maritime historian David Cordingly noted, this method of representing seawater persisted with only slight modifications for the following five centuries.[8] We see it, for instance, in the intricately designed seals of the southern English Cinque Ports, as well as in the blue-green swirls that represented the sea in dozens of illuminated medieval manuscripts.

As did ancient maritime writing, the earliest visual representations of the sea tended to focus on the human experience, on shipping and seafaring, rather than on the sea as a subject in its own right. The much-reproduced sixteenth-century woodcut of HMS *Ark Royal* is a typical example, in which the sea is reduced to a horizontal grey plinth on which rides the proud pennanted flagship of the Armada campaign. But in the following two centuries a growing distinction arose (within the European context) between 'maritime art', which

focused on elements of shipping and seamanship, and 'marine art', in which the sea itself took centre stage, although, as will be seen, the categories were often blurred.

One the earliest depictions of the sea in European art: a detail from the Bayeux Tapestry, *c.* 1077, in which the English Channel appears as a series of undulating lines.

In 1805 the London-based father-and-son maritime painters Dominic and John Thomas Serres published an instruction manual, the *Liber Nauticus*, in which they laid down the basis of 'proficiency in drawing Marine subjects'. The manual was distinguished by a dual focus on the visual components of ships and shipping and the compositional elements of the sea itself; as explained in the preface, addressed to 'the Amateurs of Marine Drawing', the book featured examples of 'all the different kinds of Ships and vessels; which are made still farther interesting and amusing by the introduction, in the background, of scenes taken from nature'.[9] Those scenes appeared in the many illustrations, engraved from the work of both father and son, that were included as painterly object lessons, from studies of seas in various states – 'A light Breeze with Breakers'; 'A making Tide, rolling on a Beach with a strong Breeze'; 'A brisk Gale' – to detailed outlines of a ship's manoeuvres in various states of sea, whether 'casting', 'veering', or 'scudding', the latter referring to the movement of a ship carried along by a storm:

> as a ship flies with amazing rapidity through the water, whenever this expedient is put in practice, it is never attempted in a contrary wind, unless when her condition renders her incapable of sustaining the mutual efforts of the wind and waves any longer on her side, without being exposed to the most imminent danger.[10]

Woodcut of the flagship *Ark Royal*, *c.* 1587, in which the sea appears as a horizontal grey plinth.

As was clear from the array of nautical detail that featured in the narrative, maritime painting had long been the domain of the knowledgeable, often pedantic enthusiast, by whom the smallest factual error would not go unnoticed. Early in his short career, the nineteenth-century painter George Chambers, who

had gone to sea in 1813, at the age of ten, and possessed first-hand knowledge of ships and seamanship, received a series of letters from an unusually exacting patron: Admiral Sir George Mundy had commissioned a pair of large-scale sea pictures from Chambers, and in one letter, concerning the depiction of the capture of a French frigate, Mundy praised the artist's drawing of 'the ship, of the sea and of the land', but took him to task over his apparent mishandling of the sails:

Why is the foretopsail not set? – no man of war loosens her foretopsail as a signal for sailing – unless she is in charge of a convoy and at anchor. The jib should be eased a third in on the jib boom – it will look more ship-shape in blowing weather. The mizzen or spanker need not be loose, which will shew an ensign at the mizzen peak, which you should hoist, and blow upwards like your jack at the foretopmast head – these alterations be so good as to make, and then I will pronounce it perfect.[11]

In another letter Mundy detailed how he had found fault with Chambers's depiction of smoke over water, advising him to travel to the naval town of Portsmouth on 'the first saluting day, and view it both close and distant, and I am quite confident you will see and correct the error immediately, and it will serve you ever after.'[12]

But it was just such a literal-minded approach to depiction – as championed by the *Liber Nauticus* – that would be swept away in the following decades by the painterly innovations of J.M.W. Turner, whose artistic vision was dominated by the sea, from the first oil painting he ever exhibited in London, *Fishermen at Sea* (1796), to his last, the reworked *Wreck Buoy*, which he sent to the Royal Academy in 1849, at the age of 74. Over the course of his five-decade career, the visual power of Turner's sea paintings would overturn established pictorial rules and conventions, creating 'an entirely new maritime aesthetic', in the words of Christine Riding, who cites a later treatise on marine painting by the American artist Edward Moran, which noted the extent of Turner's visual licence when it came to depicting coastal scenes:

> I once took a lot of Turner's engravings of views on the English coast, and went with them, as nearly as I could judge, to the exact spots from which they must have been taken – at Hastings, Dover and other south-eastern points – going out in a boat and rowing about until I found the right place. Well, the result settled all doubts as to his accuracy. He is very inaccurate – wilfully so . . . his knowledge of the forms of land and sea and cloud was so thorough that he could do pretty much as he pleased with them, and yet keep within the bounds of naturalness.[13]

The original images were most likely from George Cooke's *Picturesque Views of the Southern Coast of England* (1814–26), a popular album of etched and engraved versions of Turner's seaside views. Many of them offered boat's-eye views of coastal towns, sometimes from quite rough waters, as in *Brighthelmston, Sussex* (1824), in which a ferry boat is propelled towards Brighton's newly built Chain Pier by a small but powerful storm wave, the composition offering a visual foreshadowing of Susan Hiller's crashing-wave postcards.

The sea in motion afforded Turner the greatest scope for painterly expression, beginning with the naturalistic seascapes that he produced early in his career. *Dutch Boats in a Gale* (1801), an early masterpiece, had been commissioned by the Duke of Bridgewater as a companion to *A Rising Gale* (1672), a sea piece by the Dutch master Willem van de Velde the Younger, who was known in Britain as 'the most eminent Sea-Painter that did, or perhaps ever will, exist'.[14] Turner, characteristically, interpreted the commission as a direct challenge, and set about outdoing the Dutch artist with an impressively composed canvas that was filled with dramatic contrast, featuring the brilliant white peaks of a storm-whipped sea moving beneath a wall of threatening cloud. Turner, who was paid the enormous sum of £250 for the work, was aware that this episode constituted

a turning point in his career, and he later recounted how the audacious painting had 'launched my Boat at once'.[15]

But Turner's full expressiveness would emerge in his later canvases, in which waves, mist, light and cloud swirl together in the near abstract configurations that led the essayist William Hazlitt to declare that Turner 'painted pictures of nothing, and very like'.[16] Some of his late masterpieces, such as *Snow Storm – Steam-boat off a Harbour's Mouth* (1842), the lengthy subtitle of which included the claim that 'The Author was in this Storm in the Night the "Ariel" left Harwich', were mocked by newspaper critics, with reviewers complaining that they couldn't see anything through the driving rain and snow. *Snow Storm* was the origin of one of the best-known episodes in the Turner legend, after the artist declared that he did not paint it to be understood, but simply to show what such a storm was like: 'I got the sailors to lash me to the mast to observe it,' he said. 'I was lashed for four hours, and I did not expect to escape, but

J.M.W. Turner, *Whalers*, c. 1845, oil paint on canvas. One of a series of four paintings inspired by Thomas Beale's *Natural History of the Sperm Whale* (1839). In this canvas, the whale has just fatally attacked an approaching whaleboat.

I felt bound to record it if I did.'[17] Whether true or not (and there is no evidence that a ship of that name ever left Harwich during a storm) the story became a ready image of the recklessly committed artist for whom life is less important than creative expression. The now-forgotten Victorian sea painter Henry Moore (not the sculptor) borrowed the trope in relation to one of his own paintings, *Rough Weather in the Mediterranean* (1874), which he started during a winter cruise to Egypt aboard a schooner belonging to a wealthy patron. The ship ran into a storm off Alexandria, where, according to Moore's biographer, Frank Maclean, the artist had an easel and chair secured to the rail of the bridge: 'there a "weathercloth" was also erected, behind which the artist and his friend (who held the box of colours) ducked whenever a very heavy shower of spray swept over the vessel. In this somewhat perilous eyrie Moore painted with all his might.'[18]

Like Turner, Moore spent as much time at sea as he could, and most of his Royal Academy exhibits were large-scale sea pictures, although none of them achieved the success of Turner's great canvases, such as *The Fighting Temeraire* (1838), the critical highlight of the Royal Academy exhibition of 1839, and now one of Britain's best-loved works of art (appearing on the back of the new £20 note, issued in early 2020). The painting depicts one of the most famous vessels of the age, a veteran warship, being towed to a breaker's yard in Rotherhithe by a steam-powered tugboat. The 98-gun *Temeraire* had played a distin-guished role in the Battle of Trafalgar in 1805, but by 1838 she had been sold off by the Admiralty for scrap. Turner shows the condemned gunship near the end of her final journey up the Thames estuary; here was the age of sail giving way to the age of steam, and the image went on to become a poignant emblem of the waning of Britain's naval power.

In the spring of 1855 Turner's champion, John Ruskin, visited a number of British coastal towns while undertaking background research for his book, *The Harbours of England* (1856). The heart of the volume was an annotated album of twelve engraved reproductions of Turner's studies of English

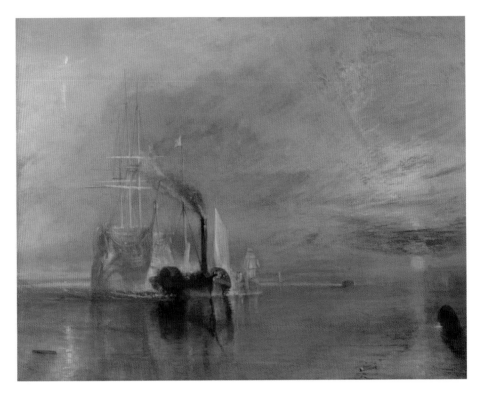

ports, for which Ruskin wrote a sequence of discursive essays in which he praised the seafaring life in idealized terms. 'Take it all in all,' he observed, 'a Ship of the Line is the most honourable thing that man, as a gregarious animal, has ever produced.'[19] The book was also a continuation of Ruskin's reflections on Turner as a sea-painter, which he had begun to formulate in *Modern Painters* (1843), and was a subject on which he could rhapsodize for England. Here is his account of Turner's more-than-visual understanding of the sea:

Turner's *The Fighting Temeraire*, 1838, oil on canvas, depicts a veteran warship being towed to a breaker's yard by a steam-powered tugboat. The 98-gun *Temeraire* had played a distinguished role in the Battle of Trafalgar, but by 1838 she had been sold for scrap.

He was assured of another fact, namely, that the *Sea* was a thing that broke to pieces. The sea up to that time had been generally regarded by painters as a liquidly composed, level-seeking, consistent thing, with a smooth surface, rising to a water-mark on sides of ships; in which ships were scientifically to be embedded, and wetted, up to said water-mark, and to remain dry above the same. But Turner found during his Southern Coast tour that the sea was *not* this: that it was, on the contrary, a very incalculable and unhorizontal thing, setting its 'water-mark' sometimes on the highest heavens, as well as on sides of ships; – very breakable into pieces; half of a wave separable from the other half, and on the instant carriageable miles inland; – not in any wise limiting itself to a state of apparent liquidity, but now striking like a steel gauntlet, and now becoming a cloud, and vanishing, no eye could tell whither; one moment a flint cave, the next a marble pillar, the next a mere fleece thickening the thundery rain . . . so stayed by him, for ever, the Image of the Sea.[20]

The influence of Turner's 'Image of the Sea' on British art cannot be overstated, and it also extended across the Channel, where a number of French painters emulated the powerful abstract realism of his later work, especially his atmospheric sea pictures. Gustave Courbet, for example, whose preoccupation with coastal scenes evolved during summers spent in Normandy during the 1850s and '60s, developed a decidedly Turneresque pictorial vocabulary for his unpeopled views of the sea and sky that he characterized as *paysages de mer* ('sea landscapes'). He painted on the beach, *en plein air*, and enjoyed considerable commercial success with a sequence of canvases that, he claimed, were knocked off in a couple of hours apiece (although he suffered a reversal when his dealer went bankrupt in 1868, owing '25,000 francs for my 25 seascapes of Trouville, plus 5,000 I lent him in cash').[21]

Courbet's seascapes fell into two major groups, 'marines' and 'waves', and were often titled accordingly. The marine series, in which several dozen canvases survive, was produced around 1865–6, and usually depicted still, calm seas at low tide, under cloud-filled skies; horizontal bands of paint, laid on with the palette knife, hinted at abstraction while being held firmly in place by the ever-present horizon line. The wave series, by contrast, focused on the motif of the cresting wave, the violent churning of which Ruskin had earlier characterized as the

greatest challenge in art. As museum curator Charlotte Eyerman observes, Courbet demonstrated 'a profound understanding that waves are like liquid sculpture, created at the intersection of water, air, and land', and his close-up wave paintings succeeded in capturing the monumental tension between the restless movement and energy of the sea, and its apparent solidity and weight.[22]

Gustave Courbet, *The Sea*, c. 1865, oil on canvas. Courbet became a near-obsessive sea painter, producing dozens of views of the Normandy shoreline during the 1860s and '70s.

The author Guy de Maupassant offered an insight into Courbet's painting method, which he observed during a visit to the artist's holiday retreat in 1869, when Courbet was working on one of his most successful sea paintings, *Stormy Sea (The Wave)* (1870):

Gustave Courbet, *The Waterspout*, 1870, oil on canvas, in which Courbet adds a meteorological slant to his familiar 'wave' motif.

In a huge, empty room, a fat, dirty, greasy man was slapping white paint on a blank canvas with a kitchen knife. From time to time he would press his face against the window and look out at the storm. The sea came so close that it seemed to batter the house and completely envelope it in its foam and roar. The salty water beat against the windowpanes like hail, and ran down the walls. On his mantelpiece was a bottle of cider next to a half-filled glass. Now and then, Courbet would take a few swigs, and then return to his work. This work became *The Wave*, and caused quite a sensation around the world.[23]

Apart from the glass of Normandy cider, the description closely matched the working practice of the American painter Winslow Homer, who developed a similarly abiding interest in the sea, spending the last two decades of his life painting seascapes at Prouts Neck, Maine, a rugged stretch of Atlantic coast about 30 km (*c.* 20 mi.) south of Portland. His studio, now a National Historic Landmark, was a remodelled carriage house overlooking the sea, although he also had a portable painting shed constructed on runners to allow him closer access to the water, even in severe weather. In a raging northeaster, when it would be otherwise impossible to paint outdoors, he would have the shed moved onto the rocks at Eastern Point, where, according to an early biographer, 'installing himself in this snug shelter, he could place himself in the position that commanded his subject, and work as long as the light and other conditions were favorable'.[24] It was there that Homer created some of his

Winslow Homer, *Cannon Rock*, 1895, oil on canvas. Homer spent twenty years painting seascapes on the Maine coast, in a studio overlooking Cannon Rock, a formation named for the booming sound of the surf breaking at its base, as well as for its distinctive shape.

Winslow Homer, *Northeaster*, 1895, oil on canvas. The unusually low perspective places the viewer directly in the line of the incoming wave.

best-known works, including *Northeaster* and *Cannon Rock* (both 1895, and both now in New York's Metropolitan Museum of Art). *Northeaster*'s unusually low perspective places the viewer directly in the line of the oncoming waves, and like Courbet's earlier great wave paintings, Homer's late masterpieces succeeded in conveying the overwhelming power and limitless energy of the sea.

Beyond the European and American contexts, wave motifs had long been a feature of Japanese art, and *The Great Wave off Kanagawa* (*c.* 1830), a colour woodblock print by Katsushika Hokusai, is now one of the world's most reproduced artworks. It depicts a large cresting wave about to engulf three *oshiokuri* – high-speed fishing skiffs – in the waters off the southern Kantō region of Japan. In the distance, a vulnerable-looking Mount Fuji, the national symbol of Japan, can be glimpsed through the trough of the breaker, affording an unusual perspective of an isolated mainland feature glimpsed from a vessel out at sea.

The French art critic Edmond de Goncourt, who did much to establish Hokusai's reputation in Europe, wrote in 1896 that 'the design for *The Wave* is a sort of deified version of the sea, made by a painter who lived in religious terror of the overwhelming sea surrounding his country on all sides', and compared the breaking wave to a predator's claw descending on the helpless boats.[25] As the sea moves in a powerful leftward direction, the wave itself curls back to the right, with bubbles of sea-spray appearing to fall as snow on the distant Mount Fuji.

Katsushika Hokusai, *The Great Wave off Kanagawa*, from *Thirty-six Views of Mount Fuji, c.* 1830. This hand-coloured woodcut has become one of the world's most familiar and widely reproduced images.

In recent years, Hokusai's plunging wave has often been misread as a tsunami, but although seismic sea waves (such as the one that caused the Fukushima nuclear disaster in March 2011) are common occurrences across coastal Japan, Hokusai's print does not depict a tsunami approaching the shore, but rather an exceptionally large storm wave breaking out at sea, with a smaller wave peak rising in front of it, suggestive of the sea's wind-blown procession of waves and troughs.[26]

There were many artists who made the sea their main preoccupation, and it is not the purpose of this book to list

Ogata Kōrin, *Rough Waves*, c. 1704–9. This energetic view of a stormy sea was painted in ink and gold leaf on a two-panel Japanese folding screen.

them, even if space allowed.[27] But it would be an oversight not to mention the St Ives School, a loose grouping of European artists who made their way to west Cornwall during the early to mid-twentieth century, and whose interest in abstraction was mirrored by the shapes, forms and colours of this littoral landscape. The proximity of the blue Atlantic shaped much of their work, from the folk-art fishing scenes of Alfred Wallis, to the later saltwater abstractions of Terry Frost or Patrick Heron. For the sculptor Barbara Hepworth, who moved to St Ives in 1939, the sea became a recurring motif: the bronze *Sea Form (Porthmeor)* (1958), for example, writhes like a conch shell wrested from the waves, like something foraged on one of Hepworth's daily coastal walks. In fact the sinuous half-painted

form *Pendour* (1947), now in Washington, DC, was carved from a length of plane wood picked up on the beach at Pendour Cove, 9 km (6 mi.) west along the coast path from St Ives.

Albert Pinkham Ryder, *Under a Cloud*, *c.* 1900.

'The visual thrust was straight out to sea', as Hepworth characterized her cliff-top studio (now a museum), and in one of her final interviews, she described watching the Atlantic rollers surging onto the shore at St Ives, sculpting the beach below, while she worked in her garden overlooking the waves. 'The sea, a flat diminishing plane, held within itself the capacity to radiate an infinitude of blues, greys, greens and even pinks of strange hues,' she recalled, 'the incoming and receding tides made strange and wonderful calligraphy on the pale granite sand.'[28]

Sculpted waves: detail of a cold-cast resin sculpture by Mitchell House, depicting the rescue of the crew of the Norwegian brig *Ispolen* by the Sheringham Lifeboat, Norfolk, during a storm in January 1897.

The sea on film

The advent of photography in the mid-nineteenth century directed a new form of documentary attention to the sea. Like painters, still photographers were drawn to the challenge of representing moving water, and the subject often featured in early photography magazines, such as *Amateur Photographer* (est. 1884), whose editor Francis Mortimer documented his growing enthusiasm for 'big wave hunting'; an article published in 1903 described the elaborate, if reckless, method he devised for photographing storm waves, which involved a companion holding him steady at the end of a rope, as he approached the incoming breakers in his oilskins and sou'wester, camera held tightly in both hands.[29] On other occasions he tethered himself

to the mast of a pitching boat, Turner-like, in pursuit of what he termed 'wavescapes' – pictorialist visions of foaming seas – with his box camera protected beneath layers of oilcloth and Vaseline.

Seas and shorelines became such familiar subjects for early photography that a new profession arose in their train, as portrait photographers plied their trade on crowded Victorian and Edwardian beaches. Equipped with a mobile darkroom, a seaside photographer could develop beach portraits for sitters within minutes, either as ambrotypes (glass-plate positives) or ferrotypes (direct positive images on enamelled iron), creating a new genre of near-instantaneous image – the holiday snap, with the beach in the background – that reinforced the idea of the seaside as a site of instant gratification.[30]

When moving pictures arrived in the 1890s, the sea proved just as alluring to early film-makers as it had to still photographers.

The French painter and pioneer photographer Gustave Le Gray transformed seascape photography into an art form, producing many dramatic images, such as *The Great Wave, Sète*, 1857. This view of the French Mediterranean coast was a composite print, with the stormy clouds derived from one negative, the crashing sea from another.

The British Film Institute's Victorian collection includes a sizeable subsection devoted to 'Sea Wave Films', many of which are available to view online. As the BFI's website observes, 'the "sea wave" genre might be one of the more surprising genres to come out of early film. But for Victorian audiences there was something hypnotic about these compact studies of movement.'[31] The short films, most of which date from the late 1890s, offer fixed-camera shots of seas in various states, such as *Rough Sea at Dover* (1895) or the gentler *Incoming Tide* (1898), filmed on the beach at Worthing, West Sussex. These sequences exploited the possibilities of the new technology of moving film, offering mesmerizing meditations on movement itself, in contrast to the 'Rough Sea' stills which were already established as a photographic genre. The rush and retreat of sea waves suited the

Rhyl, photographed from its since-demolished pier, in an albumen silver print from glass negative, by Francis Bedford, *c.* 1870s.

fixed-camera technique, as the movement played out before the static box with no need for adjustment in focus.

One of the earliest short films for cinema, Louis Lumière's *Barque sortant du port* (A Small Boat Leaving the Harbour, 1895) also employed a single fixed take, in which a group of women and children watch from a jetty as three men take to the sea in a rowing boat, which bobs and sways in the incoming waves. The boat pushes further out, encountering an increasingly agitated sea. 'With the coming of a particularly large wave, they lurch to the left and – the screen goes blank. Their fate remains unknown,' as the film historian Erika Balsom describes it, observing that the film shows both the ocean and the cinema, 'united by inhuman animus and a penchant for flux', conspiring to dislodge humanity from its pedestal.[32]

The following century saw a wide range of seas invoked by film-makers, from seas of adventure and risk, as in John Ford's *The Seas Beneath* (1931) or Michael Curtiz's maritime noir, *Sea Wolf* (1941), to the supernatural science-gone-wrong sea, from B-movie pictures such as *It Came from Beneath the Sea* (dir. Robert Gordon, 1955) or *Attack of the Crab Monsters* (dir. Roger

Louis Lumière's *Barque sortant du port* (1895), a 46-second short film in which a rowing boat ventures into a rough sea.

Corman, 1957), to more recent big-budget spectaculars such as James Cameron's *The Abyss* (1989), or Jon Turteltaub's *The Meg* (2018), in which a hitherto unknown layer of deep ocean is discovered beneath a thermocline in the Mariana Trench, wherein lurks a 25-metre (82 ft) megalodon, a species of giant shark that has, in reality, been extinct for nearly 4 million years.

Such 'creature features' continue to borrow heavily from the underwater documentary style pioneered by Jacques Cousteau in the 1950s and '60s, notably in his Oscar-winning feature *The Silent World* (*Le Monde du silence*, 1956), co-directed with Louis Malle, and its bathyspheric follow-up, *World Without Sun* (*Le Monde sans soleil*, 1964), which chronicled Cousteau's Continental Shelf Station (the so-called 'Conshelf') project, the first serious attempt to create an undersea research environment in which 'oceanauts' could live and work on the seabed for extended periods of time. Cousteau had been a naval instructor during the Second World War, and had experimented with a portable underwater breathing device that he co-patented in 1946 as the Aqua-lung: the world's first commercially available scuba equipment. He went on to develop a waterproof camera case, which, combined with the Aqua-lung, allowed him to spend ever-longer periods filming underwater, and in 1952 he shot the first ever deep-water colour film footage while diving in the Red Sea. For the next half-century Cousteau sailed the world on the *Calypso*, a decommissioned naval minesweeper which he had converted into a floating laboratory and film studio; from there he produced his series of award-winning books, films and television programmes about the undersea realm, while promoting his increasingly ambitious Conshelf project.

There had in fact been an earlier attempt at establishing an underwater photographic studio. In the late 1930s, the British inventor and film-maker John Ernest Williamson devised the 'photosphere', a diving chamber attached to a flexible metal breathing tube, in which he had pioneered submarine photography and film-making in a number of locations around the world. In 1939 he moored his bathysphere – the *Jules Verne* – on

the sea floor off the port of Nassau, in the Bahamas, and (with the permission of the Bahamian authorities) began issuing stamps from 'the world's first undersea post-office' as a revenue-raising exercise. Williamson's post office featured its own franking machine, with which he added the now collectable postmark 'Sea Floor, Bahamas' to pictorial envelopes that he sold as souvenirs to the many visitors who joined him in his photosphere. The coming of the war may have restricted Williamson's activities, but the concept of undersea living continued to fascinate researchers, from Jacques Cousteau to the architects of the Sealab experiments of the late 1960s, which were abandoned after the nine-person Sealab III sprang a fatal leak at 190 m (610 ft) off the coast of California.

Post-war maritime films often portrayed the sea as an enemy (or the haunt of an enemy), as in *The Cruel Sea* (dir. Charles Frend, 1953) or *The Enemy Below* (dir. Dick Powell, 1957), an adversarial attitude that has persisted into more recent portrayals, such as *All Is Lost* (dir. J. C. Chandor, 2013), in which Robert Redford's yacht collides with a stray shipping container in the Indian Ocean, leaving him battling for survival amid the relentless waves. Passing container ships fail to spot Redford's

A souvenir envelope from John Williamson's submarine post office, franked 'Sea Floor, Bahamas, May 6, 1940'. A report in the *Miami News* on 31 December 1939 described the installation as 'the mecca for Nassau's visitors who make trips in the photosphere to see for themselves the extraordinary variety of undersea life which exists in the Bahamas and the beautiful undersea coral and flora formations'.

damaged vessel, which drifts through the sea of commerce, of 'container capitalism', that is as indifferent to his human plight as the physical sea itself. Chandor is not the only director to have envisioned the sea as a kind of pitiless screen: 'It is simply a surface and a horizon, and it is timeless,' observes the film historian Tony Thomas, who contends that any story filmed at sea will be altered by its commanding presence.[33] The temperament of the sea itself – unlike the studio tanks in which most early nautical movies, such as *Moby Dick* (dir. Lloyd Bacon, 1930) or *The Sea Wolf* (dir. Michael Curtiz, 1941), were shot – shaped the filming process as much as the end product, and the practical difficulties experienced during the making of sea-shot films such as *Jaws* (1975) or the second *Moby Dick* (1956) were noted in an earlier chapter. The Polish director Jerzy Skolimowski recalled his time on the film *The Lightship* (1985), a slice of saltwater gothic shot in the sea off Denmark, a location dictated by the scarcity of working lightships in the era of automatic buoys. Skolimowski confirmed that every cautionary tale about filming at sea turned out to be true, not least the effect on the actors' temperaments. As with *Jaws*, the

The crew of the U.S. Navy's experimental underwater habitat, *Sealab 1*, spent eleven days submerged off Bermuda in July 1964, before a tropical storm forced them to resurface. The vessel is now on display at the Museum of Man in the Sea, Florida.

combination of claustrophobia and seasickness had a terrible effect on both cast and crew; Skolimowski described the whole experience as 'nightmarish', and pointed out that it was a rare director who ever made *two* films at sea.[34]

Sea music

In the summer of 1905 Claude Debussy rented a room at the Grand Hotel, Eastbourne (a genteel English seaside resort), where he stayed throughout July and August. He was in the process of divorcing his first wife, Lilly, while completing the symphonic sketches that would become one of his best-known works: *La Mer*. Debussy found Eastbourne 'a charming, peaceful spot; the sea unfurls itself with an utterly British correctness . . . what a place for working in!'[35] Proximity to the sea seemed to sharpen his focus, for during the earlier stages of composition Debussy had resisted visiting the coast, preferring to draw on half-remembered, half-imagined impressions of the sea from childhood trips to the Côte d'Azur and Brittany. 'The sea fascinates me to the point of paralysing my creative faculties,' he explained in an interview in 1914, in which he elaborated on his contradictory relationship with the subject of his most famous work: 'I've never been able to write a page of music under the direct, immediate impression of this great, blue sphinx, and my "symphonic triptych" *La Mer* was entirely composed in Paris.'[36] The claim was untrue, for in fact Debussy moved around a great deal during the two years he spent working on *La Mer*, including a three-week sojourn to the Channel Islands in 1904, and a summer spent on the coast at Dieppe, where he had intended to finish the piece, but found himself struggling with the orchestration, which, he wrote, was 'as tumultuous and varied as the sea itself'.[37] The structure was also subject to intense revision throughout this period, with Debussy eventually settling on the three named sections, or movements: *De l'aube à midi sur la mer* ('From dawn to midday on the sea'), which conforms to the traditional opening movement of a symphony; *Jeux de vagues* ('Play of the waves'), a formally complex sequence of interrelated

Debussy chose a detail from Hokusai's *Great Wave off Kanagawa* for the published score of *La Mer* (1905), declaring that he was pleased with 'the shiny outlines they've given the waves on the cover'.

CLAUDE DEBUSSY

LA MER

motifs; and *Dialogue du vent et de la mer* ('Dialogue of the wind
and the sea'), which loosely takes the form of a rondo. As music-
ologist Caroline Potter observed, 'Debussy avoids monotony by
using a multitude of water figurations that could be classified as
musical onomatopoeia: they evoke the sensation of swaying
movement of waves and suggest the pitter-patter of falling
droplets of spray', while avoiding the already clichéd arpeggiated
triads, as used by Wagner and Schubert, as a standard musical
means of evoking the play of water.[38]

The premiere of *La Mer* was given on 15 October 1905 in
Paris, the programme for which likened *La Mer* to a painting,
observing that the orchestral effects of the piece were achieved
through 'a palette of sounds and by skilful brushstrokes designed
to convey in gradations of rare and brilliant colours the play of
light and shade and the *chiaroscuro* of the ever-changing
seascape'.[39] This first performance was not a triumph, however,
and Debussy was particularly stung by a newspaper review that
appeared the following day, in which the prominent critic Pierre
Lalo, complained that 'I do not hear, I do not see, I do not smell
the sea' in *La Mer*. Debussy wrote to Lalo, insisting that 'I love
the sea and I've listened to it with the passionate respect it
deserves. If I've been inaccurate in taking down what it dictated
to me, that is no concern of yours or mine.'[40]

But it would not be long before *La Mer* became established
as an orchestral favourite, a soothing musical equivalent to one
of Boudin's or Monet's seascapes. It also served to place
Eastbourne on the musical map; in 1911 the Brighton-born
composer Frank Bridge went there to complete his own marine
suite, *The Sea*, which in turn inspired Arnold Bax's symphonic
poem *Tintagel* (1919), which was written during a sojourn on
the Cornish coast. The piece was intended to offer 'a tonal
impression of the castle-crowned cliff of Tintagel', and more
especially of the long expanse of the Atlantic as seen from the
cliffs.[41] Much of Bax's work was written in response to the sea;
the opening of his Symphony No. 4 (1930), known as the 'Sea
Symphony', was, according to Bax, 'inspired by a rough sea at
flood-tide on a sunny day', while in the opening of his

Symphony No. 6 (1935), the listener can 'perceive a seascape through the presence of a melodic archetype that resembles the depiction of the ocean at the beginning of *The Garden of Fand* (another of Bax's sea-pieces).[42] For Bax, the sea was more than a Celticist motif; it was a part of his musical life, and he spent a great deal of time in a cottage on the coast of Donegal, communing with the sea:

> In winter I would often linger at my window, too fascinated in watching the implacable fury of the Atlantic in a south-westerly storm to sit down to work. At one end of the little Glen Bay was a wilderness of tumbled black rocks . . . and upon this grim escarpment the breakers thundered and crashed, flinging up, as from a volcano, towering clouds of dazzling foam which would be hurled inland by the gale to put out the fires in the cottage hearths . . . the savagery of the sea was at times nearly incredible.[43]

Bax later claimed that his deathbed vision would be of 'the still, brooding, dove-grey mystery of the Atlantic at twilight', a vision that in the end did not materialize, Bax dying unexpectedly during a visit to the city of Cork.

The sea was of comparable thematic importance to the Welsh composer Grace Williams, who was born in the coastal town of Barry in 1906, her parents having 'had the good sense to set up home on the coast of Glamorgan', as she wrote in the dedication to her best-known score, *Sea Sketches* (1944).[44] Her friend and fellow student at the Royal College of Music, Elizabeth Maconchy, had recognized this aspect of Williams's musical character early in their friendship, writing in a letter sent in January 1934: 'I hope you've had some decent weather for your holiday, and been able to be by the sea a bit. I think you belong to the sea.'[45]

Williams composed a number of sea-themed pieces throughout her career, but only the *Sea Sketches* have endured. The piece, a suite of five movements for string orchestra – 'High Wind'; 'Sailing Song'; 'Channel Sirens'; 'Breakers'; and 'Calm

Sea in Summer' – paint 'vivid pictures of the sea in various moods', as Williams's first biographer, Malcolm Boyd, characterized it: 'The gusts blowing in "High Wind" and the rolling of the tide in "Breakers" are powerfully suggested; "Channel Sirens" is a particularly striking evocation, with its desolate, repetitious fog-horn sounding on violas and cellos and its misty chords on the upper strings.'[46]

A calm coastal view: John Frederick Kensett, *Eaton's Neck, Long Island*, 1872, oil on canvas.

In 1945, shortly after completing the work, Williams wrote that 'I don't want to stay in London – I just long to get home and live in comfort by the sea,' a wish that she soon realized by relocating to coastal Wales where, as she later recalled, she walked for an hour by the sea every day.[47] After *Sea Sketches* was recorded in 1970 by the English Chamber Orchestra ('it was sometimes used by the BBC as fill up between programmes – or as background music for sea stories'), Williams began to develop a keen interest in new music-making technology, describing in 1972 her intention to take a course in electronic music in order to explore new ways of responding to the sounds of the sea:

I think the idea may have grown out of a recent awareness – more than ever before – of the sounds & rhythms of the sea; but I wouldn't ever want to <u>distort</u> those sounds if I fed them into a synthesizer (or whatever it's called) – & I doubt if the most authentic stereo reproduction could ever convey the open-air-space which surrounds the sea sounds: but if that were possible – then superimposing viola or alto flute or distant, disembodied voice might be exciting.[48]

Williams, who by then was in her late sixties, never did go on to create her imagined electronic seascapes, but subsequent generations of musicians have successfully done so, such as British electronic band Morcheeba's 'The Sea', from their album *Big Calm* (1998), with its dreamy invocation of the sounds of a shoreline, or the Brighton-based indie outfit British Sea Power, the title of whose debut album was borrowed from a classic work of maritime scholarship, Desmond Wettern's *The Decline of British Seapower* (1982), which argued for the retention of a strong standing navy in the light of post-war reduction. Politicians and the public, Wettern lamented, think that naval officers spend their time 'in a round of cocktail parties with a bit of gentle cruising in between', although the Falklands War,

Commemorative postcard, *c.* 1920, showing the British Grand Fleet under steam. The Grand Fleet was assembled in 1914 to counter the German High Seas Fleet during the First World War.

'Britain Prepared'   By permission of H.M. Admiralty

THE GRAND FLEET UNDER WAY   [Reference No. 1

which broke out the week after the book was published, went on to make Wettern's argument for him.[49]

Sea shanties

The calm coastal impressionism of *La Mer* and its imitators was a world away from the kinds of music that were made and performed at sea, such as the ballads and shanties (or 'chanteys') that arose from life on board merchant ships. The intricate taxonomy of sea shanties speaks to the history of the sea as a taskscape, in which songs and sounds arose in concert with specific manual occupations. Most shanties were sung without musical accompaniment by whichever members of a ship's crew were engaged in a particular task, with the three main families of shanty – hauling shanties, heaving shanties and special-occasion shanties – distinguished primarily by occupational type. A task determined its associated song's particular rhythm, cadence, structure and duration.

The main types of work song (hauling and heaving) gave rise to notably task-specific arrangements. Hauling shanties, for example, were of four types: (1) 'long-drag' shanties, sung when hoisting the largest and heaviest sails on the ship, as well as during long spells of rowing; (2) 'short-drag' shanties, sung when hoisting higher, smaller sails and sheets, such as topgallants and royals; (3) 'furling sail' shanties, sung when furling and folding sails aloft; and (4) 'hand-over-hand' shanties, which were short-haul songs sung when setting jibs or other small sails.[50] The best-known sea shanty of all – 'What shall we do with a drunken sailor?' – began life as a hand-over-hand shanty to be sung at the halyard, its memorable pace and rhythm established by the nature of the work. Such repetitive tasks could sometimes take many hours, giving rise to open-ended task songs such as 'Blow the Man Down', first recorded in the 1860s, in which verses and choruses could be altered and improvised to an indefinite length.

'Blow the Man Down' featured in Eugene O'Neill's one-act drama, *The Moon of the Caribbees* (1918), one of a quartet of short

SATURDAY NIGHT AT SEA.

plays set on a British tramp steamer, which were later combined into a full-length drama, *s. s. Glencairn* (1924), and filmed during the Second World War as *The Long Voyage Home* (dir. John Ford, 1940). In the scene, the crew sit idle on the deck one evening as the vessel lies at anchor off an unnamed Caribbean island. Bored, they ask one of the older sailors, Driscoll, for a song ('give us a chanty, Drisc, one all av us knows'), to which the Irishman retorts that seamen no longer know the words to the old songs – 'Ye've heard the names av chanties but divil a note av the tune or a loine av the words do ye know. There's hardly a rale deep-water sailor left on the seas, more's the pity.'[51] Then one of the seamen says, 'Give us "Blow the Man Down", we all know some of that', and the shanty starts up with the whole crew joining in unaccompanied:

> DRISCOLL Come in then, all av ye. [*He sings:*]
>   As I was a-roamin' down Paradise Street –
> ALL Wa-a-ay, blow the man down!

DRISCOLL As I was a-roamin' down Paradise Street –
ALL Give us some time to blow the man down!

CHORUS
Blow the man down, boys, oh, blow the man down!
Wa-a-ay, blow the man down!
As I was a-roamin' down Paradise Street –
Give us some time to blow the man down![52]

As O'Neill's soon-to-be-drunken sailors demonstrate, this was a well-known hauling shanty, the lyrics of which refer, subversively, to a man-o'-war being literally blown down into the water when a sudden gale catches it with its topsails fully set. The role of Driscoll as the lead 'chantyman' reflects the call-and-response nature of the shanty tradition, with the chantyman democratically appointed each time by the rest of the singing crew.

Heaving shanties, by contrast, were sung when working the capstan or windlass, or when weighing anchor, loading cargo or manning the pumps, which were also tasks that could go on for many hours. Unlike hauling shanties, heaving shanties were sometimes accompanied by a fiddler or accordionist, who would stand on the capstan as the men pushed or pulled in time to the music. This longer kind of shanty was characterized by improvisable ballad-like lyrics with choruses featuring two or more lines, as in the popular eighteenth-century ballad 'Spanish Ladies', or the weighing-anchor song 'Outward Bound' (first recorded in the 1860s), with its regular heaving rhythm and its cadence of buoyant farewell:

Solo: From the West Indies docks I bid adieu
To lovely Sal, and charming Sue;
Our ship's unmoored, our sails unfurled,
We are bound to plough the watery world.
Chorus: *For we are outward bound;*
Hurrah! We are outward bound![53]

Historic coastal defence battery, Lundy Island, Bristol Channel.

By the end of the nineteenth century, the advent of steam power, along with the mechanization of most shipboard tasks, had brought about the demise of the shanty. Singing was no longer a necessary accompaniment to the drudgery of work at sea. But over the course of the following century shanties would be revived as objects of antiquarian interest, and today most shanties are sung on land rather than on the decks of seagoing vessels, where they offer a fleeting musical connection to an otherwise vanished seafaring culture, a way of life preserved in half-remembered songs and stories that tell of the terrible grandeur of the boundless sea, where life stops and the unknown begins.

Afterword: Future Seas

Where are your monuments, your battles, martyrs?
Where is your tribal memory? Sirs,
in that grey vault. The sea. The sea
has locked them up. The sea is History.
Derek Walcott, 'The Sea Is History' (1978)

There is an Argentinian folk tale that tells of a young oyster fisherman from Bahía Blanca, who farmed his own area of the bay. The young man knew that starfish preyed on oysters, so whenever he caught one in his net he would tear it in half and throw it back into the sea. But what he didn't know was that many species of starfish don't die when ripped in half; instead, each part regenerates itself, so for every starfish he caught and rent, two grew back in the sea. Over time he began to wonder why so many starfish – and so few oysters – seemed to inhabit his waters, until he was told about starfish regeneration by an older and wiser fisherman. From that day on, or so the story goes, he left the starfish alone.[1]

It's a story of innocence and experience, and of the law of unintended consequences, but it is also a prescient object lesson in environmental protection. Our seas have been heavily depleted by decades of overfishing, and today a third of the world's fish stocks are currently being overfished, and consequently pushed beyond their reproductive limits. The Atlantic bluefin tuna, for example, has been fished almost to extinction, with an 80 per cent population decline since the 1970s, a fate shared with many other sought-after food species such as sea bass, monkfish and orange roughy. By the end of the twentieth century, industrial fishing by commercial fleets had reduced global populations of large ocean fish to around 10 per cent of their pre-industrial numbers. Today's hi-tech fishing vessels land a fraction of the annual catch of nineteenth-century

A blue starfish
(*Linckia laevigata*)
resting on hard corals,
Great Barrier Reef,
Australia.

trawlers; though, even then, declining catches had already been observed, with commercially exploited fish beginning to fall in numbers throughout British and Irish waters from the 1820s onwards, following the widespread adoption of bottom trawling or 'dragging', a practice that involves towing a large fishing net (a 'beam trawl') along or just above the sea floor in an effort to maximize the catch. At the same time, the introduction of the glass float, replacing the heavier wooden float, had led to a marked increase in the size of fishing nets.

The term 'overfishing' was coined in 1850, in a Scottish Fishery Board report which noted that, 'by the statements of Fishermen generally, it appears that the Boats are almost everywhere obliged to go further from land than formerly': 'and hence it is assumed either that the Fish have changed their runs on account of the Fishing that has been carried on, or that the Fishing grounds near the shore have been over-fished'.[2]

Two parliamentary commissions were appointed to look into the claims, and their detailed reports, published in 1866 and 1885, concluded that the observable depletion of inshore fish stocks had largely resulted from an increase in deep-sea trawling. Whitefish species such as cod and ling were particularly impacted by the practice, their populations suffering notable decline from the 1840s onwards, but even the famously abundant herring, the 'silver darlings' of a hundred folk songs, had reduced in number since the advent of industrialized trawling. In W. B. Yeats's 'The Meditation of the Old Fisherman' (1886), the poem's narrator laments that:

> The herring are not in the tides as they were of old;
> My sorrow! for many a creak gave the creel in the cart
> That carried the take to Sligo town to be sold.[3]

In recent years, bottom trawling has rendered the seabed of the southern North Sea almost lifeless through the use of weighted nets that pulverize reefs and oysterbeds, but the practice long pre-dated the nineteenth and twentieth centuries. Trawling technology had been the subject of complaints as early as the 1370s, when a petition was presented to the English parliament calling for the prohibition of a 'subtlety contrived instrument called the *wondyrchoum*', a 3-metre-wide (10 ft) wooden beam trawl with a weighted net of so small a mesh that 'no manner of fish, however small, entering within it can pass out and is compelled to remain therein and be taken':

Fishing trawler coming in to port, Walberswick, Suffolk.

by means of which instrument the fishermen aforesaid take
so great abundance of small fish, that they know not what to
do with them, but feed and fatten the pigs with them, to the
great damage of the whole commons of the kingdom, and
the destruction of the fisheries in like places.[4]

It's unsettling to recognize that the principal negative impacts
of commercial trawling – damage to the marine environment,
the use of fine meshed nets to catch juvenile fish and indiscrim-
inate fishing to supply animal feed – had already been identified
more than six hundred years ago.

Today, legislation restricting the use of beam trawls is in
effect across the world, but it only covers local territorial waters.
Out in the open sea, beyond national jurisdictions, bottom-
trawling remains both widespread and unregulated. A report
published in 2004 by the World Conservation Union concluded
that the practice was 'highly destructive to the biodiversity
associated with seamounts and deep-sea coral ecosystems, and
likely to pose significant risks to this biodiversity, including the
risk of species extinction'. A motion to ban beam trawling in all
waters was put before the United Nations in 2006, but was
vetoed by member states concerned by the economic impact on
their fisheries.[5]

Today there are as many as 100,000 fishing trawlers working
at sea at any one time, and the damage to marine life caused by
commercial fishing cannot be overstated. The conclusion is stark:
unless we collectively change the way we manage and exploit
ocean species, and learn to respect the seas as working ecosystems,
then by the end of the present century, the consumption of wild-
caught seafood may have become nothing but a distant memory.[6]

Overfishing is not the only threat to marine life. A post-
mortem conducted on an adult female killer whale (*Orcinus orca*)
that washed up on the Scottish coast in November 2016 found
high levels of chemicals stored in her blubber, notably toxic
PCBs: polychlorinated biphenyls, human-made chemicals known
to cause adverse health effects, including infertility, in humans
and wildlife. The use of PCBs has been banned in most countries

'Can you see the sea monster?': a seaside telescope with a container ship on the horizon, the new monsters of the open seas.

since the 1970s, but millions of tonnes of them had already made their way into the world's oceans, where apex predators such as killer whales have proved particularly susceptible to their effects. Chemical concentrations increase as they make their way up the food chain, from crabs and molluscs to small fish, then bigger fish, and finally into the bodies of top marine predators such as sharks and killer whales. Along with other legacy pollutants such as heavy metals, pesticides, pharmaceuticals and microplastics, the prevalence of these long-lived chemicals jeopardizes the future health of all marine ecosystems.

Life on Earth has always depended on the oceans' natural life-support system, with deep waters playing a vital role in the detoxification of the planet. In shallow waters, plankton and other microscopic organisms excrete, shed their shells and eventually die, sending all this organic material down to the depths in the form of 'marine snow': a continuous shower of detritus, some of which can take several weeks to make its way to the bottom.

Marine snow – the term was coined in the 1930s by the undersea explorer William Beebe – transfers a significant amount of energy from the light-rich photic zone to the dark, aphotic zone below, either as food that can be eaten on its way down, or as nutrient-rich sea-floor sediments, a transfer known as the biological pump. The snow is also responsible for cleaning seawater as it descends, dragging pollutants such as sewage, oil and other contaminants down with it, burying them in the seabed for thousands of years. But the list of contaminants now includes ever-increasing volumes of microplastics – tiny pieces of plastic less than 5 mm in diameter. Millions of tonnes of plastic, rubber, polystyrene and fibreglass waste (at least 8 million tonnes per year, according to the World Economic Forum) continue to find their way into the ocean, where they are broken down into microplastics and nanoplastics, most of which are invisible to the eye, some of which are not: a distressing array of marine animals, from seabirds to beaked whales, have been found starved to death, their bellies filled with plastic objects apparently mistaken for food.

Early 20th-century Lifebuoy Soap advertisement, invoking the life-saving associations of 'natural' hygiene and cleanliness.

But the invisible micro-particles are just as deadly; broken down by sunlight and wave action, they are consumed along with natural nutrients by zooplankton at the bottom of the food chain. Like any other ingested chemical, such micro- (and even smaller nano-) particles increase in concentration and toxicity the further up the food chain they travel. Hundreds of species of marine creature are now known to have consumed this growth-stunting plastic soup, including 70 per cent of deep-sea fish, while wind- and wave-blown particles and fibres have been found in every corner of the ocean, whether in deep-sea sediments dredged from miles below the surface (including in the Mariana Trench, the deepest spot in the ocean), or frozen into vast areas of Arctic sea ice, or pasted onto remote coral reefs in the Indian and Pacific Oceans. Once they are in deep water, microplastics tend to congregate in hotspots, swept by deep-sea currents into the seabed equivalent of the notorious Pacific 'garbage patches' visible on the surface. In certain areas of ocean, such as the Yellow Sea off South Korea, there are some 10 billion pieces of litter per square kilometre of water, three-quarters of which is ubiquitous plastic, derived from a wide range of anthropogenic sources.[7]

Coral reefs are among the many organisms impacted by the plastic confetti, suffering energy depletion from the task of clearing debris from their polyps, the soft, anemone-like parts of the organism used for feeding. Corals are already under stress from ocean warming, with coral bleaching events increasing all around the world. When water temperatures rise by more than one or two degrees Celsius above normal for prolonged periods, coral polyps respond by expelling the tiny algae that live within their tissues, causing them to turn completely white. These symbiotic algae (known as zooxanthellae) help corals to photosynthesize food, and are what give reefs their dazzling colour palettes. Once coral has been bleached of its algae, it is in danger of starving to death, as happened to nearly half of Australia's Great Barrier Reef during a sustained bleaching event in 2016–17, in the wake of an unusually strong El Niño (the warm phase of a climate cycle that can drive up sea and air temperatures across much of the tropical Pacific). The catastrophic bleaching has impacted the survival not only of the coral itself, but of thousands of reef-dwelling species, such as molluscs and small fish, along with the groupers, sharks and octopi that feed on them. Reefs have long faced a host of human hazards, from souvenir hunters, ships' anchors, dredging and sewage run-off, but bleaching is the most serious threat so far. Reef systems – the world's largest living structures – have proved particularly vulnerable to rising temperatures, with the loss of nearly half the world's coral reefs over the past thirty years. According to a study published in *Nature Scientific Reports* in 2016, if warming continues at its present rate, more than 90 per cent of remaining reefs will be dead by the end of this century.[8]

And then there's climate change. As temperatures rise, carbon dioxide (CO_2) is changing the oceans chemically as well as thermally. The seas are absorbing ever-increasing quantities of carbon dioxide from the atmosphere, while continuing to store more than one-quarter of all the CO_2 ever emitted. The absorbed gas reacts with seawater to produce a weak carbonic acid: before the industrial era, alkaline chemicals dissolved in river water washed into the sea, neutralizing any natural build-up of acid, but carbon dioxide emissions have increased to such a degree that rivers can no longer keep up. The result is the gradual acidification of the oceans, firstly at the surface and then, through mixing, down into the depths. The seas today are around 26 per cent more acidic than they were two centuries ago.

Ocean acidification has become one of the greatest threats to marine life, since the extra acidity dissolves the calcium carbonate that is needed to form the shells and skeletons of many sea creatures, including coral. Some 250 million years ago, the catastrophic extinction event known as 'the great dying' saw the oceans acidified by intense volcanic activity, resulting in the rapid extinction of most sea life. Could prehistory be repeating itself on our watch? 'Coral bleaching used to be what kept me awake at night; now it is ocean acidification,' wrote Callum Roberts in his landmark study *Ocean of Life: How Our Seas Are Changing* (2012):

> Warming seas have devastated reefs worldwide by weakening the sunlit coalition of corals and algae. Acidification is a punch in the gut to reefs that are already on their knees. It is chilling to think that, within the space of 100 years, humanity could reverse a process of coral reef formation that has flourished since the end of the ice age.[9]

Roberts's conclusions are stark, but it is clear that the twentieth century was a disaster for the natural world, which continues to face unsustainable stress from rampant human activity. The oceanic forecast is not a reassuring one: alongside the decline in coral, seagrass meadows have been disappearing at an increasing rate, due largely to the flush of nutrients from agricultural run-off, while mangrove swamps are also vanishing, cut down for coastal development or to make way for industrial shrimp farms. Coastal dead zones – areas of water where nothing can live – have increased in size and number over the past few decades, the result of nitrogen and phosphate run-off from fertilized fields, which stimulates the growth of enormous algal blooms. As the algae dies and rots, it absorbs all available

A major coral bleaching event at the Great Barrier Reef, Australia, in 2016–17, caused by rising sea temperatures. Less than 0.1 per cent of the ocean is made up of coral reefs, yet they are the world's richest shallow-water ecosystems,

supporting more than 25 per cent of all marine species. By the end of this century, if sea temperatures continue to rise, 90 per cent of all the oceans' coral will have undergone bleaching. oxygen from the water, thereby rendering it lifeless. Overfishing, meanwhile, has spelled ruin for fisheries around the world, from the collapse of the American sardine industry in the 1950s – once the largest fishery in the western hemisphere – to the Grand Banks disaster of 1992, in which the Atlantic northwest cod fishery collapsed after decades of relentless exploitation. Add in the steady expansion of coastal and marine infrastructure associated with fish farming, container shipping, desalination, undersea cabling, offshore oil and gas drilling, and, most recently, seabed mining, and the scale of the pressure on our seas seems overwhelming. To add to these woes, a new political–industrial phenomenon, named 'seabed grabbing', is of growing concern to oceanographers. Under the terms of the United Nations Convention on the Law of the Sea (UNCLOS), maritime nations can apply to extend their continental shelves by laying claim to areas of seabed – including its rich mineral deposits – beyond the traditional 370-kilometre (200 nautical mi.) economic limits. Since the first claim was made, by Russia, in 2001, dozens more countries have made similar submissions, including small island states that are seeking to become large ocean-mining states; the Cook Islands in the South Pacific, for example, has laid claim to an area of seabed that is nearly 2,000 times its land surface, while Australia has recently secured the rights to an additional 2.5 million square kilometres (965,255 sq. mi.) of seabed around the uninhabited sub-Antarctic territory of Heard and McDonald Islands. Put together, the claims amount to over 37 million square kilometres (14,285,780 sq. mi.) of seabed, a new mining frontier that is more than twice the size of Russia.[10]

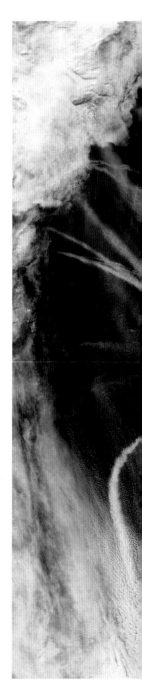

As Rachel Carson noted in the introduction to her best-selling study *The Sea Around Us* (1951), 'it is a curious situation that the sea, from which life first arose, should now be threatened by the activities of one form of that life.'[11] Carson's generation was the first to recognize that human activity was leaving its mark on every part of the seas and oceans, from the poles to the tropics, and that the resulting physical, chemical and biological stressors were already unsustainable. Half a century later, and the 'curious situation' is now a crisis. Hardly a day goes by without a depressing news story concerning the sea, usually connected to human-made pollution turning up in some unexpected place, or some new and urgent threat to marine life from rampant commercial activity.

It is customary for books on environmental subjects to end on hopeful rather than disheartened notes, but in truth it's hard to be hopeful about the long-term future of our seas. There are, of course, plenty of determined campaigners working to raise

Power at sea: the Thorntonbank Wind Farm, an offshore installation some 30 km (19 mi.) off the Belgian coast.

Pages 200–201. Ship tracks over the northern Pacific Ocean, captured by NASA's Terra satellite, 2009. Seeded from ships' engine exhaust, these are the maritime equivalent of aircraft contrails.

A herd of walrus hauled up onto sparse summer sea ice, Bering Sea.

awareness of the multiple challenges facing the oceans, and there are plenty of inspiring publications that end with check-lists of beneficial actions that can be taken to help conserve the seas: Alex Rogers's *The Deep* (2019), for example, ends with a six-step summary of individual actions ranging from considerate shopping to environmental advocacy, while Callum Roberts's *Ocean of Life* ends with insightful guidelines for consuming 'Seafood with a Clear Conscience'.[12] In 2009, one of the world's most tireless marine activists, Sylvia Earle, founded Mission Blue, an international campaigning organization that works to establish a network of marine protected areas (known as 'hope spots'), with the ultimate aim of declaring 30 per cent of the ocean protected by 2030. At the time of writing (February 2021), Mission Blue had established more than 120 marine reserva-tions across the world: an impressive achievement, yet they

add up to a fraction of the total area of protected national parks on land.

Perhaps a more radical approach is needed, such as the idea put forward by the historian David Abulafia, at the end of his magisterial history of the oceans, *The Boundless Sea* (2019). Rather than depending on existing protective legislation, he argues, the United Nations Educational, Scientific and Cultural Organization (UNESCO) should simply designate the ocean as a single, vast World Heritage Site, subject to the same conservation rules

The mix of commercial and leisure craft that dot the English Channel illustrates some of the many competing demands placed on coastal waters.

that apply to comparable sites on land. Sustainable fishing would continue, for example, but destructive seabed mining would not. Perhaps then, with the ocean as a recognized global conservation zone, humanity might begin to learn to value the seas and their fragile ecosystems, learning from the mistakes of the past, just as the oysterman in the story learned, in the end, to let the starfish be.

TIMELINE

793   Lindisfarne (Holy Island) suffers the first recorded Viking sea raid

1275   Wreck Act (3 Edw. 1) rules that if anyone ('a Man, a Dog, or a Cat') survived a shipwreck, then the ship's cargo could not be considered legal wreck

1376   Petition to ban beam trawling presented to the English parliament

1497–9   Vasco da Gama sails beyond the Cape of Good Hope, opening up a new trade route to India

1519–22   Ferdinand Magellan sails to the Pacific, his ship completing the first circumnavigation of the world under the command of Juan Sebastián Elcano after Magellan's death in the Philippines

1584   Publication of Lucas Janszoon Waghenaer's *Spieghel der zeevaerdt* (Mariner's Mirror), the first modern nautical chart

1595   Publication of Captain John Davis's *Seaman's Secrets*, which contained the first description of the backstaff, a navigational instrument that measures the altitude of the sun by the projection of a shadow

1611   Shakespeare's final play, *The Tempest*, first performed

1620   First navigable submarine constructed by Dutch inventor Cornelis Drebbel

1664   Launch of HMS *Experiment*, William Petty's innovative double-hulled sloop (lost in a storm, 1665)

1725   Italian naturalist Luigi Ferdinando Marsigli publishes *The Physical History of the Sea*, a founding work of oceanography

1736   Reverend William Clarke and his wife pioneer the summer beach holiday at Brighton (then known as Brighthelmstone)

1753   Wreck Act (26 Geo. 11, c. 19) redefined 'stealing, taking away or destroying of any goods or Merchandize belonging to a ship which shall be wrecked' as a capital offence

1805   Publication of John Davis's *The Post-Captain; or, the Wooden Walls Well Manned*, regarded as the first naval novel. Publication of Dominic and John Thomas Serres's marine painter's manual, the *Liber Nauticus*

1840   3 January: Sir James Clark Ross takes the first modern deep-sea sounding at latitude 27°s, longitude 17°w

1855   Matthew Fontaine Maury publishes *The Physical Geography of the Sea*, the founding text of modern oceanography

1856   Paris Declaration Respecting Maritime Law outlaws privateering

1859   Wreck of the passenger ship *Royal Charter* during a storm off the coast of Anglesey. The disaster leads to the establishment of the first gale warning service by the newly founded Met Office

1869   From the deck of HMS *Porcupine*, Wyville Thomson dredges living creatures from 4,400 m (2,400 fathoms) in the North Atlantic, overturning the concept of a lifeless azoic zone below 300 fathoms

1870   Jules Verne publishes his submarine novel *Vingt mille lieues sous les mers* (Twenty Thousand Leagues Under the Seas), first in magazine instalments, then in book form

1873   3 January: the crew of HMS *Challenger* make their first soundings, seen as the birth of modern oceanography

1885   Prince Albert I of Monaco outfits four yachts as oceanographic research vessels, going on to establish pioneering oceanographic research institutes in Monaco and Paris

1902   International Council for the Exploration of the Sea founded in Copenhagen

1903   Scripps Institution of Oceanography founded in San Diego, relocating to La Jolla, California, two years later

1905   Debussy's *La Mer* premieres in Paris

1930   Woods Hole Oceanographic Institution founded in Massachusetts; the largest independent oceanographic research institution in the U.S.

1944   Grace Williams composes *Sea Sketches*, a suite of five movements for string orchestra

1946   The International Whaling Commission (IWC) established.

1951   Marine biologist Rachel Carson publishes her best-selling book *The Sea Around Us*, which won a U.S. National Book Award. HMS *Challenger II* makes soundings of the ocean floor; the Challenger Deep is identified as the deepest part of the ocean

1959   First International Oceanographic Congress held in New York

1960   Jacques Piccard and Donald Walsh descend nearly 11 km
        (7 mi.) to the bottom of the Challenger Deep, in the
        bathyscaphe *Trieste*

1964   U.S. Navy *Sealab 1* is occupied by four crew members for
        eleven days, nearly 60 m (192 ft) below the surface, to test the
        plausibility of underwater living

1966   Second International Oceanographic Congress held in
        Moscow

1975   *Jaws* (dir. Steven Spielberg)

1978   16 March: supertanker *Amoco Cadiz* runs aground off Brittany,
        releasing nearly 220,000 tonnes of crude oil into the sea

1982   Conclusion of the United Nations Convention on the Law of
        the Sea (UNCLOS)

1985   1 September: the wreck of the *Titanic* discovered at a depth
        of 3.8 km (12,500 ft), some 600 km (370 mi.) off the coast of
        Newfoundland

1986   The International Whaling Commission issues a worldwide
        ban on commercial whaling

1989   24 March: supertanker *Exxon Valdez* strikes a reef in Prince
        William Sound, Alaska, spilling 37,000 tonnes of crude oil into
        the sea

1992   28,800 'Friendly Floatee' plastic bath toys released into the
        mid-Pacific during a January storm

1994   The United Nations Convention on the Law of the Sea
        (UNCLOS) comes into force, overseeing commercial and
        environmental aspects of the High Seas

1995   Merchant Shipping Act consolidated and updated British
        maritime legislation, replacing the Merchant Shipping Act
        1894

1998   UN International Year of the Ocean (YOTO)

2004   Boxing Day tsunami, Indian Ocean

2007   Antony Gormley's tidal installation, *Another Place*, becomes a
        permanent fixture at Crosby Beach, Merseyside

2012   Film director James Cameron becomes the third person to
        reach the Challenger Deep, 11 km (7 mi.) below the Pacific
        Ocean, in the submersible *Deepsea Challenger*

2013   MS *Nordic Orion* becomes the first freighter to complete the
        Northwest Passage

2019   World's largest container ship, the 400-metre-long (1,300 ft)
        MSC Gülsün, built in South Korea, but registered in Panama,
        is launched. It has a cargo capacity of nearly 24,000 6-metre
        (20 ft) shipping containers

REFERENCES

Introduction: 'The Sea Is Like Music'

1 Carl Jung, *Memories, Dreams, Reflections*, trans. Richard and Clara Winston (London, 1963), p. 339.
2 Ibid., pp. 339–40.
3 Victor Hugo, *The Toilers of the Sea*, trans. James Hogarth (New York, 2002), p. 251.
4 See Rose George, *Deep Sea and Foreign Going: Inside Shipping, the Invisible Industry that Brings You 90% of Everything* (London, 2013), pp. 3–4.
5 W. H. Smyth, *The Sailor's Word-Book: An Alphabetical Digest of Nautical Terms*, revd edn (London, 1867), p. 599. 'Ocean' is defined in the same book as 'the whole body of salt water which encompasses the globe, except the collection of inland seas, lakes, and rivers'.
6 See International Hydrographic Organization, *Limits of Oceans and Seas*, 3rd edn (Monaco, 1953), available at https://epic.awi.de, accessed 10 October 2020.
7 Jules Verne, *Twenty Thousand Leagues Under the Seas*, trans. William Butcher (Oxford, 1998); curiously, David Coward's 2017 translation for Penguin Classics opted for the singular 'Sea'. It is worth noting that the '20,000 leagues' (*c.* 80,000 km) of Verne's title refers to the horizontal distance travelled beneath the sea(s) rather than any vertical descent: the greatest depth mentioned in the book is 4 leagues (16 km), some 5 km deeper than the Challenger Deep, the deepest spot in any of the world's oceans.
8 John Smith, *The Seaman's Grammar* (London, 1653), cited in J. H. Parry, 'Sailors' English', *Cambridge Journal*, 2 (1948–9), p. 669.
9 *Old English Poems*, trans. Cosette Faust and Stith Thompson (Chicago, IL, 1918), p. 71.
10 Cited in Martin Caiger-Smith, *Antony Gormley* (London, 2017), p. 90.
11 See Trevor Murphy, *Pliny the Elder's Natural History: The Empire in the Encyclopedia* (Oxford, 2004), p. 172.

12 Caiger-Smith, *Antony Gormley*, p. 91.

1 Shorelines

1 Jonathan Raban, *Passage to Juneau: A Sea and Its Meanings* (London, 1999), p. 213; *The Poems of Matthew Arnold*, ed. Miriam Allott and Robert H. Super (Oxford, 1986), p. 136. Arnold composed a companion poem, 'Calais Sands', written from the perspective of the opposite shore, in which the poet strains to see across the Channel to England; *The Poems of Matthew Arnold*, pp. 134–5.
2 *The Poems of Matthew Arnold*, p. xvii; p. 61.
3 Ibid., p. xvii.
4 Ian McEwan, *On Chesil Beach* (London, 2007), p. 5.
5 Ibid., p. 19.
6 Robert Macfarlane, *The Wild Places* (London, 2007), p. 243.
7 Robert Macfarlane, *Landmarks* (London, 2015), pp. 165–76.
8 Greenville Collins, *Great Britain's Coasting Pilot* (London, 1693), p. 16.
9 Stephen Crane, 'The Open Boat', in *The Oxford Book of Short Stories*, ed. V. S. Pritchett (Oxford, 1981), pp. 187–8.
10 Collins, *Great Britain's Coasting Pilot*, p. i.
11 Ibid., p. 7.
12 Cited in Cathryn J. Pearce, *Cornish Wrecking, 1700–1860: Reality and Popular Myth* (Woodbridge, 2010), p. 142.
13 Cited in Michael Craton, *A History of the Bahamas* (Ontario, 1986), p. 167.
14 Cited in Pearce, *Cornish Wrecking, 1700–1860*, p. 7.
15 Daniel Defoe, *A Tour Through the Whole Island of Great Britain*, ed. Pat Rogers (London, 1986), pp. 235–6.
16 Cited in Pearce, *Cornish Wrecking, 1700–1860*, p. 1.
17 See K. R. Howe, *Where the Waves Fall: A New South Sea Islands History from First Settlement to Colonial Rule* (London, 1984), p. 103.
18 Joseph Conrad, *Lord Jim: A Tale*, ed. Jacques Berthoud (Oxford, 2002), p. 5.
19 Cited in Howe, *Where the Waves Fall*, pp. 104–7.
20 Alain Corbin, *The Lure of the Sea: The Discovery of the Seaside in the Western World, 1750–1840*, trans. Jocelyn Phelps (Cambridge, 1994), p. 78.
21 Ibid., p. 78. Cited more fully in Fred Gray, *Designing the Seaside: Architecture, Society and Nature* (London, 2006), p. 17.
22 Gray, *Designing the Seaside*, p. 17.
23 Cited ibid., p. 46. Michelet's claim is undermined by the fact that there had been a number of earlier treatises promoting the healthfulness of seawater, such as John Floyer's *History of Cold Bathing* (1702).

24 Cited in Gray, *Designing the Seaside*, p. 21. Published studies such as Jan Ingenhousz's 'On the Degree of Salubrity of the Common Air at Sea', *Philosophical Transactions*, LXX (1780), offered medical backing to the belief in the healthfulness of sea air.

25 Cited in Gray, *Designing the Seaside*, p. 147.

26 Tobias Smollett, *The Expedition of Humphry Clinker*, ed. Lewis M. Knapp (Oxford, 1984), pp. 178–9.

27 Ibid., p. 179.

28 John Bigsby, *Sea-Side Manual for Invalids and Bathers* (1841), cited in Gray, *Designing the Seaside*, p. 29.

29 Gray, *Designing the Seaside*, p. 25.

30 Jane Austen, *Persuasion*, ed. James Kinsley (Oxford, 2004), p. 85; it was a scene that John Fowles later built on in *The French Lieutenant's Woman* (1969), a novel sustained by its central image: 'a woman stands at the end of a deserted quay [the Cob at Lyme Regis] and stares out to sea': John Fowles, 'Notes on an Unfinished Novel', in Malcolm Bradbury, ed., *The Novel Today: Contemporary Writers on Modern Fiction* (Manchester, 1977), p. 136.

31 Jane Austen, *Sanditon*, ed. Kathryn Sutherland (Oxford, 2019), pp. 8, 13.

32 Janet Todd, '"Lady Susan", "The Watsons" and "Sanditon"', in *The Cambridge Companion to Jane Austen*, ed. Edward Copeland and Juliet McMaster, 2nd edn (Cambridge, 2011), p. 95.

33 Cited in Mavis Batey, *Jane Austen and the English Landscape* (London, 1996), p. 126.

34 Ibid.

35 Edmund Gosse, *Father and Son: A Study of Two Temperaments*, ed. Michael Newton (Oxford, 2004), pp. 59–60.

36 Edmund Gosse, *The Life of Algernon Charles Swinburne* (London, 1917), p. 8. In the book Gosse goes on to claim that Swinburne 'loved the sea as it was never loved before, even by an Englishman' (p. 83).

37 Algernon Swinburne, *Lesbia Brandon*, ed. Randolph Hughes (London, 1952), p. 7.

38 Ibid., pp. 9, 15.

39 Herman Melville, *Moby-Dick; or, The Whale*, ed. Tony Tanner (Oxford, 1988), pp. 1–2.

40 *The Works of John Ruskin*, ed. E. T. Cook and A. Wedderburn, 39 vols (London, 1902–12), XXXV, p. 78.

41 In *The Works of John Ruskin*, III, p. 494. *The Harbours of England* (London, 1856) was an extended commentary on a collection of twelve engravings of Turner's views of English ports.

2 The Science of the Sea

1 Curtis Ebbesmeyer and Eric Scigliano, *Flotsametrics and the Floating World* (New York, 2009), pp. 83–4.
2 Matthew Fontaine Maury, *The Physical Geography of the Sea*, new edn (New York, 1858), pp. 29–30.
3 In Edgar Allan Poe, *Selected Tales*, ed. Julian Symons (Oxford, 1980), p. 13. See also Ebbesmeyer and Scigliano, *Flotsametrics and the Floating World*, pp. 55–6.
4 Poe, *Selected Tales*, p. 17.
5 Charles Dickens and Wilkie Collins, 'A Message from the Sea', in *The Christmas Stories*, ed. Ruth Glancy (London, 1996), p. 381.
6 See Anthea Trodd, 'Messages in Bottles and Collins's Seafaring Man', *Studies in English Literature, 1500–1900*, XLI/4 (2001), pp. 751–64.
7 Ebbesmeyer and Scigliano, *Flotsametrics and the Floating World*, pp. 56–7. Elizabeth I created an official position of 'Uncorker of Ocean Bottles' and, believing that some bottles might contain secrets from British spies, ordered the execution of anyone else caught opening them. In 1876, on the remote Scottish island of St Kilda, journalist John Sands and a party of marooned Austrian sailors deployed two messages seeking rescue from the island. The messages, enclosed in cocoa tins attached to sheep bladders, made landfall in Orkney within nine days and in Ross-Shire two weeks later. The letters were forwarded to the admiralty, and a short while later HMS *Jackal* arrived at St Kilda to rescue the petitioners. Since that time, sending a 'St Kilda mail boat' has become something of a ritual for island visitors, who launch containers to ride the Gulf Stream eastwards to the Scottish mainland, and even on as far as Scandinavia.
8 Aristotle, *Meteorologica*, trans. E. W. Webster (Oxford, 1931), 356a–357b.
9 Ted McCormick, *William Petty and the Ambitions of Political Arithmetic* (Oxford, 2009), p. 151.
10 Ibid., p. 155.
11 Susan Schlee, *A History of Oceanography: The Edge of an Unfamiliar World* (London, 1975), p. 17.
12 Thomas R. Anderson and Tony Rice, 'Deserts on the Sea Floor: Edward Forbes and his Azoic Hypothesis for a Lifeless Deep Ocean', *Endeavour*, XXX/4 (2006), p. 132.
13 C. P. Idyll, ed., *The Science of the Sea: A History of Oceanography* (London, 1970), p. v.
14 Cited in *At Sea With the Scientifics: The 'Challenger' Letters of Joseph Matkin*, ed. Philip F. Rehbock (Honolulu, HI, 1992), p. 40.
15 See Helen M. Rozwadowski, *Fathoming the Ocean: The Discovery and Exploration of the Deep Sea* (Cambridge, MA, 2005), p. 191.

16 Schlee, *A History of Oceanography*, p. 16.

17 Rozwadowski, *Fathoming the Ocean*, p. 5.

18 W. H. Smyth, *The Sailor's Word-Book: An Alphabetical Digest of Nautical Terms*, revd edn (London, 1867), p. 172.

19 *Coleridge's Notebooks: A Selection*, ed. Seamus Perry (Oxford, 2002), p. 60.

20 Ralph Abercromby, *Seas and Skies in Many Latitudes; or, Wanderings in Search of Weather* (London, 1888), p. 160.

21 Tristan Gooley, *How to Read Water: Clues, Signs and Patterns from Puddles to the Sea* (London, 2016), p. 253.

22 Ibid., p. 263.

23 Hugh Aldersey-Williams, *Tide: The Science and Lore of the Greatest Force on Earth* (London, 2016), p. 21.

24 See ibid., pp. xxxi–ii.

25 Ibid., p. 331.

26 Edgar Allan Poe, 'A Descent into the Maelström', in *The Science Fiction of Edgar Allan Poe*, ed. Harold Beaver (London, 1976), pp. 72–88; Jules Verne, *Twenty Thousand Leagues Under the Seas*, trans. Lewis Mercier (London, 1872), pp. 300–302; A. S. Byatt, *The Biographer's Tale* (London, 2000), pp. 246–8. Byatt's elusive character, the biographer Scholes Destry-Scholes, was last seen leaving a Norwegian fishing port on a small boat. The abandoned craft was found in the water a week later, 'not far from the Moskenes Current, more famous by its fifteenth-century name, the Maelstrøm'.

27 Poe, 'A Descent into the Maelström', pp. 74–5.

28 Verne, *Twenty Thousand Leagues*, p. 301. The novel ends with the *Nautilus* caught in the deadly whirlpool, 'whirled round with giddy speed', before apparently disappearing into the 'terrible gulf' (p. 303).

29 Cited in Esmond Wright, *Franklin of Philadelphia* (Cambridge, MA, 1986), p. 58.

30 Cited in Carl van Doren, *Benjamin Franklin* (New York, 1938), p. 522.

31 A. D. Bache, 'Lecture on the Gulf Stream, Prepared at the Request of the American Association for the Advancement of Science', *American Journal of Science and Arts*, 80 (1860), pp. 313–29; see also Hugh Richard Slotten, *Patronage, Practice, and the Culture of American Science: Alexander Dallas Bache and the U.S. Coast Survey* (Cambridge, 1994), pp. 136–9.

32 Matthew Fontaine Maury, *The Physical Geography of the Sea*, 2nd edn (New York, 1858), pp. 25, 50–51. This passage was reproduced almost word for word in Jules Verne's submarine novel, *Twenty Thousand Leagues Under the Seas* (1870).

3 Sea Life

1 Ernest Hemingway, *The Old Man and the Sea* (New York, 1952),
   p. 81.
2 Jules Verne, *Twenty Thousand Leagues Under the Seas*, trans. Lewis
   Mercier (London, 1872), p. 11.
3 Edmund Gosse, *Father and Son: A Study of Two Temperaments*, ed.
   Michael Newton (Oxford, 2004), p. 82.
4 Ibid., p. 81.
5 Philip Henry Gosse, *A Naturalist's Rambles on the Devonshire Coast*
   (London, 1853), pp. 25–6.
6 Philip Henry Gosse, *The Aquarium: An Unveiling of the Wonders
   of the Deep Sea*, 2nd edn (London, 1856), pp. 21–2. Of course, not
   everyone shared the Victorian passion for rock pooling. John
   Ruskin recalled that, as a child in the 1830s, the sea itself was
   the main attraction: 'I never took to natural history of shells, or
   shrimps, or weeds, or jelly-fish. Pebbles? – yes, if there were any;
   merely stared all day long at the tumbling and creaming strength of
   the sea. Idiotically, it now appears to me.' *The Works of John Ruskin*,
   ed. E. T. Cook and A. Wedderburn, 39 vols (London, 1903–12),
   XXXV, p. 78.
7 Gosse, *The Aquarium*, pp. 55–6.
8 Bernd Brunner, *The Ocean at Home: An Illustrated History of the
   Aquarium* (New York, 2005), pp. 47–8.
9 Benjamin Franklin, 'Journal of a Voyage, 1726', in *The Oxford Book of
   the Sea*, ed. Jonathan Raban (Oxford, 1992), p. 100.
10 Gosse, *Father and Son*, p. 81.
11 *The True Discription of this Marueilous Straunge Fishe* (London,
   1569), https://quod.lib.umich.edu, accessed 10 September 2019.
12 Cited in Mary Colwell, 'How Jaws Misrepresented the Great
   White', www.bbc.co.uk, 9 June 2015.
13 Peter Benchley, *Jaws* (London, 1974), pp. 304–5.
14 Edgar Allan Poe, *The Narrative of Arthur Gordon Pym of Nantucket
   and Related Tales*, ed. J. Gerald Kennedy (Oxford, 1998), p. 100.
15 Julia K. Baum et al, 'Collapse and Conservation of Shark
   Populations in the Northwest Atlantic', *Science*, 299 (2003), pp.
   389–92.
16 Cited in James Honeyborne and Mark Brownlow, *Blue Planet II:
   A New World of Hidden Depths* (London, 2017), p. 290.
17 Jody Bourton, 'Giant Bizarre Deep Sea Fish Filmed in Gulf of
   Mexico', BBC Earth News, http://news.bbc.co.uk, 8 February 2010.
18 Pliny the Elder, *Natural History*, trans. H. Rackham, 10 vols
   (London, 1952), III, p. 171. According to one school of thought, the
   story of the Loch Ness monster could have arisen from oarfish
   sightings.

19 Pliny the Elder, *Natural History*, III, pp. 169–71.
20 Ibid.
21 Cited in Henry Lee, *Sea Monsters Unmasked* (London, 1883), p. 1.
22 Ibid., pp. 4–5.
23 Alfred Tennyson, *The Major Works*, ed. Adam Roberts (Oxford, 2009), p. 20.
24 Seamus Perry, *Alfred Tennyson* (Tavistock, 2005), p. 43.
25 'The Sea Raiders', in *The Short Stories of H. G. Wells* (London, 1927), pp. 189–94.
26 Verne, *Twenty Thousand Leagues*, trans. Mercier, p. 286.
27 Ibid., p. 276.
28 Victor Hugo, *Toilers of the Sea*, trans. W. Moy Thomas (London, 1911), pp. 294–5.
29 Ibid., p. 295.
30 See Yulia V. Ivashchenko and Phillip J. Clapham, 'What's the Catch? Validity of Whaling Data for Japanese Catches of Sperm Whales in the North Pacific', *Royal Society Open Science*, II/7 (2015), https://royalsocietypublishing.org, accessed 2 March 2020.
31 From George Mackay Brown, *Selected Poems: 1954–1992* (London, 1996), pp. 27–8.
32 Herman Melville, *Moby-Dick; or, The Whale*, ed. Tony Tanner (Oxford, 1988), pp. 419–20.
33 Ian McGuire, *The North Water* (London, 2016), p. 103.
34 Alastair Fothergill and Keith Scholey, *Our Planet* (London, 2019), p. 270
35 Jonny Keeling and Scott Alexander, *Seven Worlds, One Planet: Natural Wonders from Every Continent* (London, 2019), p. 176.
36 Angus Atkinson et al., 'Krill (*Euphausia superba*) Distribution Contracts Southward During Rapid Regional Warming', *Nature Climate Change*, 9 (2019), pp. 142–7. The paper concludes that 'the changing distribution is already perturbing the krill-centred food web.'

4 Exploration

 1 Stick charts were introduced to European readers in 1862, in an article in *The Nautical Magazine*, which described 'rude maps by which they [the Marshall Islanders] retain and impart knowledge regarding the direction and distances of the various groups. These maps consist of small sticks tied together in straight or curved lines, intended to represent the currents or waves to be met, while the islands are to be found at certain points where these lines meet': L. H. Gulick, 'Micronesia', *The Nautical Magazine and Naval Chronicle*, 31 (1862), p. 304.
 2 Brian Fagan, *Beyond the Blue Horizon: How the Earliest Mariners Unlocked the Secrets of the Oceans* (London, 2012), p. xiii.

3 Cited in Joseph Genz, 'Polynesian and Micronesian Navigation', in *The Oxford Encyclopedia of Maritime History*, ed. John B. Hattendorf, 4 vols (Oxford, 2007), III, pp. 144–54.

4 Cited in David Lewis, *We, the Navigators: The Ancient Art of Landfinding in the Pacific*, 2nd edn (Honolulu, HI, 1994), p. 244.

5 Ibid., p. 245.

6 Ibid., p. 164.

7 Lyall Watson, *Heaven's Breath: A Natural History of the Wind* (New York, 2019), p. 99.

8 Herodotus, *The Histories*, trans. Aubrey de Sélincourt (Harmondsworth, 1954), p. 255.

9 Thor Heyerdahl, *Early Man and the Ocean* (London, 1978), pp. 24–6.

10 Ibid., p. 36.

11 Watson, *Heaven's Breath*, p. 100.

12 *The Periplus of the Erythræan Sea: Travel and Trace in the Indian Ocean by a Merchant of the First Century*, trans. Wilfred H. Schoff (London, 1912), pp. 37–8. See also Lionel Casson, *The Periplus Maris Erythraei: Text with Introduction, Translation and Commentary* (Princeton, NJ, 1989).

13 Cited in E. H. Warmington, *The Commerce Between the Roman Empire and India*, 2nd edn (London, 1974), pp. 58, 67.

14 Pliny the Elder, *Natural History*, trans. W. H. S. Jones, 10 vols (London, 1953), VIII, p. 465.

15 Laurence Bergreen, *Over the Edge of the World: Magellan's Terrifying Circumnavigation of the Globe* (New York, 2003), p. 205.

16 Jonathan Raban, *Passage to Juneau: A Sea and Its Meanings* (London, 1999), pp. 93–5.

17 Greenville Collins, *Great Britain's Coasting Pilot* (London, 1693), p. 16.

18 Joseph Conrad, *Typhoon and Other Tales*, ed. Cedric Watts (Oxford, 1986), pp. 31–2.

19 Raban, *Passage to Juneau*, p. 98.

20 Watson, *Heaven's Breath*, p. 92.

21 *The Tempest*, I, 2, 197–201. The Strachey/*Tempest* parallels are outlined in detail in Geoffrey Bullough, *Narrative and Dramatic Sources of Shakespeare*, 8 vols (London, 1957–75), VIII, pp. 275–94. See also Alden T. Vaughan and Virginia Mason Vaughan, *Shakespeare's Caliban: A Cultural History* (Cambridge, 1991), p. 40, and their introduction to the play, in *The Tempest*, ed. Virginia Mason Vaughan and Alden T. Vaughan (London, 1999), p. 42. John Fletcher and Philip Massinger's comedy *The Sea Voyage* (1622) borrowed much of *The Tempest*'s setting and story.

22 *York Mystery Plays: A Selection in Modern Spelling*, ed. Richard Beadle and Pamela M. King (Oxford, 1995), p. 15.

23 In *The Towneley Plays*, ed. Martin Stevens and A. C. Cawley, 2 vols (Oxford, 1994), I, p. 43.

24 The term 'taskscape' first appeared in Tim Ingold, 'The Temporality of the Landscape', *World Archaeology*, XXV/2 (1993), pp. 152–74.

25 Cited in John Mack, *The Sea: A Cultural History* (London, 2011), p. 190.

26 W. H. Smyth, *The Sailor's Word-Book: An Alphabetical Digest of Nautical Terms*, revd edn (London, 1867), p. 7; Mack, *The Sea*, p. 190.

27 Cited in J. H. Parry, 'Sailors' English', *Cambridge Journal*, 2 (1948–9), p. 670; and in Mack, *The Sea*, p. 191.

28 Geoff Dyer, *Another Great Day at Sea: Life Aboard the USS George H. W. Bush* (Edinburgh, 2015), p. 35; Redmond O'Hanlon, *Trawler: A Journey through the North Atlantic* (London, 2003), p. 41.

29 Christer Westerdahl, 'Seal on Land, Elk at Sea: Notes on the Ritual Landscape at the Seaboard', *International Journal of Nautical Archaeology*, XXXIV/1 (2005), p. 4.

30 Ibid., pp. 2–8; Mack, *The Sea*, p. 189; see also Remco Knooihuizen, 'Fishing for Words: The Taboo Language of Shetland Fishermen and the Dating of Norn Language Death', *Transactions of the Philological Society*, 106 (2008), pp. 100–113.

31 O'Hanlon, *Trawler*, p. 8.

32 From an endnote in Conrad, *Typhoon and Other Tales*, p. 303.

33 Frederick Marryat, *The Phantom Ship* (London, 1839), p. 76.

34 Cited in Brian Hicks, *Ghost Ship: The Mysterious True Story of the 'Mary Celeste' and her Missing Crew* (New York, 2004), pp. 20–21.

35 John Davis, *The Post-Captain; or, the Wooden Walls Well Manned; Comprehending a View of Naval Society and Manners* (London, 1805), pp. 2, 4; Parry, 'Sailors' English', p. 661. There had been earlier novels set at sea, such as Defoe's pirate tale *The Life and Adventures of Captain Singleton* (1720), but Davis's was the first to reflect the working conditions of ship-board life.

36 Charles Dickens and Wilkie Collins, 'A Message from the Sea', in *The Christmas Stories*, ed. Ruth Glancy (London, 1996), p. 377.

37 Walter Benjamin, 'The Storyteller', in *Illuminations*, trans. Harry Zohn (London, 1973), p. 85.

38 Frederick Marryat, *Peter Simple*, new edn (London, 1888), pp. 207, 256.

39 Cited in Christopher Lloyd, *Captain Marryat and the Old Navy* (London, 1939), p. 33; see also Patrick Brantlinger, *Rule of Darkness: British Literature and Imperialism, 1830–1914* (Ithaca, NY, 1988), p. 49.

40 Marryat, *Peter Simple*, pp. 92–3.

41 In *Joseph Conrad on Fiction*, ed. Walter F. Wright (Lincoln, NE, 1964), p. 48.

42 Elizabeth Molt, 'Language', in *The Oxford Encyclopedia of Maritime History*, II, p. 318.
43 Fred Weeks et al., *SeaSpeak Training Manual: Essential English for International Maritime Use* (Oxford, 1988).
44 William Langewiesche, *The Outlaw Sea: Chaos and Crime on the World's Oceans* (London, 2005), p. 4.
45 Ibid., p. 5.
46 Conrad, *Typhoon and Other Tales*, p. 10.
47 Rose George, *Deep Sea and Foreign Going: Inside Shipping, the Invisible Industry that Brings You 90% of Everything* (London, 2013), p. 21.
48 Krista Maglen, 'Health and Health Care', in *The Oxford Encyclopedia of Maritime History*, II, pp. 135–6.
49 In Jonathan Swift, *The Complete Poems*, ed. Pat Rogers (Harmondsworth, 1983), p. 208. See also Herman Melville, *Moby-Dick*, ed. Tony Tanner (Oxford, 1988), p. 500; and Edgar Allan Poe, *The Narrative of Arthur Gordon Pym of Nantucket and Related Tales*, ed. J. Gerald Kennedy (Oxford, 1998), p. 74.

5 The Sea in Art and Music

1 In *Susan Hiller*, ed. Ann Gallagher (London, 2011), p. 30.
2 A sixty-page album of selected postcards was published to accompany the first exhibition: Susan Hiller, *Rough Sea* (Brighton, 1976).
3 *Susan Hiller*, p. 76.
4 *The Diary of Virginia Woolf*, ed. Anne Oliver Bell, 5 vols (London, 1984), V, p. 116. My thanks to Katherine Angel for the reference.
5 *The Works of John Ruskin*, ed. E. T. Cook and A. Wedderburn, 39 vols (London, 1902–12), III, p. 561.
6 *The Works of John Ruskin*, III, pp. 561–2.
7 David T. Ansted, 'The Representation of Water', *Art Journal*, 2 (1863), p. 13. An extract from the essay is included in Jonathan Raban, ed., *The Oxford Book of the Sea* (Oxford, 1992), pp. 265–7.
8 David Cordingly, *Marine Painting in England, 1700–1900* (London, 1974), p. 15.
9 Dominic and John Thomas Serres, *Liber Nauticus, and Instructor in the Art of Marine Drawing* (London, 1979), p. 1.
10 Ibid., pp. 4–12.
11 Cited in John Watkins, *Life and Career of George Chambers* (London, 1841), pp. 39–40.
12 Ibid., p. 40.
13 In Christine Riding and Richard Johns, *Turner and the Sea* (London, 2013), pp. 11, 13–14.
14 Cited ibid., p. 31.

15 Cited ibid., p. 33.
16 Cited in James Attlee, *North Sea: A Visual Anthology* (London, 2017), p. 20.
17 Cited in Riding and Johns, *Turner and the Sea*, p. 246.
18 Frank Maclean, 'Henry Moore R.A.' (London, 1905), cited in Cordingly, *Marine Painting in England*, p. 173. Moore also claimed to have spent five hours on Hastings beach during a severe storm, making preparatory sketches for an earlier painting, *Winter Gale in the Channel*, which was exhibited at the Royal Academy in 1872.
19 John Ruskin, *The Harbours of England* (London, 1856), p. 12.
20 Ibid., pp. 23–4. With its visionary paeans to the intensity of the sea, *The Harbours of England* proved a critical success, with one reviewer describing it as 'a prose poem worthy of a nation at whose throne the seas, like captive monsters, are chained and bound' (cited in *The Works of John Ruskin*, XIII, p. xxi).
21 Mary Morton and Charlotte Eyerman, *Courbet and the Modern Landscape* (Los Angeles, CA, 2006), p. 103; Jack Lindsay, *Gustave Courbet: His Life and Art* (Bath, 1973), p. 237.
22 Cited in Morton and Eyerman, *Courbet and the Modern Landscape*, p. 104.
23 Guy de Maupassant, 'La Vie d'un paysagiste', *Gil Blas*, 28 September 1886, in *Études, Chroniques et Correspondance de Guy de Maupassant*, ed. René Dumesnil (Paris, 1938), pp. 167–8.
24 William Howe Downes, *The Life and Works of Winslow Homer* (New York, 1989), p. 176.
25 Edmond de Goncourt, *Hokusai*, cited in Edward Lockspeiser, *Debussy: His Life and Mind*, 2 vols (London, 1965), II, p. 24.
26 Timothy Clark, *Hokusai's Great Wave* (London, 2011), pp. 8–9; see also Julyan H. E. Cartwright and Hisami Nakamura, 'What Kind of Wave is Hokusai's *Great Wave Off Kanagawa?*', *Notes and Records of the Royal Society*, 63 (2009), pp. 119–35.
27 There are already plenty of monographs on the subject, such as David Cordingly, *Painters of the Sea* (London, 1979), Frank B. Cockett, *Early Sea Painters, 1660–1730* (New York, 1999), or E. H. Archibald, *The Dictionary of Sea Painters of Europe and America* (New York, 1999).
28 Barbara Hepworth, *Carvings and Drawings* (London, 1952), note 4, unpaginated.
29 Francis Mortimer, 'On a Rocky Coast', *Amateur Photographer*, 38 (1903), cited in Val Williams and Karen Shepherdson, *Seaside Photographed* (London, 2019), pp. 49–50.
30 Though the influential documentary photographer Walker Evans went on to rail against 'nature' subjects, declaring that 'under no circumstances is [photography] anything ever anywhere near a

beach'. Cited in Peter C. Bunnell et al., *EW 100: Centennial Essays in Honor of Edward Weston* (Carmel, 1986), p. 48.

31 British Film Institute, 'Sea Wave Films', https://player.bfi.org.uk, accessed 2 January 2020.

32 Erika Balsom, *An Oceanic Feeling: Cinema and the Sea* (New Plymouth, NZ, 2018), pp. 17–19.

33 Tony Thomas, *The Cinema of the Sea: A Critical Survey and Filmography, 1925–1986* (London, 1988), p. 66.

34 Ibid., p. 186.

35 *Debussy Letters*, ed. François Lesure and trans. Roger Nichols (London, 1987), p. 153. In a later letter, he described Eastbourne as 'a little seaside town, as ridiculous as these sorts of places always are – too many draughts and too much music, both of which I try and avoid but I don't really know where to go!', ibid., p. 156.

36 Cited in Robert Orledge, 'Debussy the Man', in *The Cambridge Companion to Debussy*, ed. Simon Trezise (Cambridge, 2003), p. 23.

37 Cited in Simon Trezise, *Debussy: La Mer* (Cambridge, 1994), p. 14.

38 Caroline Potter, 'Debussy and Nature', in *The Cambridge Companion to Debussy*, p. 149.

39 Cited in Lockspeiser, *Debussy*, p. 16.

40 *Debussy Letters*, pp. 163–4.

41 Lewis Foreman, *Bax: A Composer and His Times* (London, 1983), p. 150.

42 Cited in Aidan J. Thomson, 'Bax's "Sea Symphony"', in *The Sea in the British Musical Imagination*, ed. Eric Saylor and Christopher M. Scheer (Woodbridge, 2015), pp. 226–7.

43 Cited ibid., p. 228.

44 Grace Williams, *Sea Sketches for String Orchestra* (Oxford, 1951), p. i.

45 Cited in Jenny Doctor, 'Afterword: Channelling the Swaying Sound of the Sea', in *The Sea in the British Musical Imagination*, p. 270.

46 Cited ibid., p. 271.

47 Ibid., p. 270.

48 Cited ibid., p. 273.

49 Desmond Wettern, *The Decline of British Seapower* (London, 1982), p. i. Large naval wars are seldom-seen affairs today, since nations with substantial navies rarely fight each other; most wars now are either civil wars or some form of asymmetrical warfare, fought on land, and often with the involvement of military aircraft. The main function of the modern navy is to exploit its control of the seaways to project power ashore.

50 See Stuart M. Frank, 'Ballads and Chanteys', in *The Oxford Encyclopedia of Maritime History*, ed. John B. Hattendorf, 4 vols (Oxford, 2007), I, pp. 246–9.

51 In Eugene O'Neill, *The Long Voyage Home: Seven Plays of the Sea* (New York, 1940), p. 9.
52 Ibid., p. 10.
53 Cited in Frank, 'Ballads and Chanteys', p. 248.

Afterword: Future Seas

 1 Story told in Colum McCann, *Fishing the Sloe-black River* (London, 1995), p. 35.
 2 *Fishery Board Annual Report*, 1850, cited in Peter Jones et al., 'Early Evidence of the Impact of Preindustrial Fishing on Fish Stocks from the Mid-West and Southeast Coastal Fisheries of Scotland in the 19th Century', ICES *Journal of Marine Science*, LXXIII/5 (2016), pp. 1410–14.
 3 In *The Collected Poems of W. B. Yeats*, ed. Richard J. Finneran (London, 1991), p. 21.
 4 Cited in Edgar J. March, *Sailing Trawlers: The Story of Deep-sea Fishing with Long Line and Trawl* (Newton Abbot, 1970), p. 33.
 5 Matthew Gianni, *High Seas Bottom Trawl Fisheries and their Impacts on the Biodiversity of Vulnerable Deep-sea Ecosystems: Options for International Action* (Gland, Switzerland, 2004), p. 2. See also 'Ban on "Brutal" Fishing Blocked', http://news.bbc.co.uk, 24 November 2006.
 6 Boris Worm et al., 'Impacts of Biodiversity Loss on Ocean Ecosystem Services', *Science*, 314 (2006), pp. 787–90.
 7 James Honeyborne and Mark Brownlow, *Blue Planet II: A New World of Hidden Depths* (London, 2017), p. 284.
 8 Ruben van Hooidonk et al., 'Local-scale Projections of Coral Reef Futures and Implications of the Paris Agreement', *Nature Scientific Reports*, 6 (2016), www.nature.com, accessed 30 April 2020.
 9 Callum Roberts, *Ocean of Life: How Our Seas Are Changing* (London, 2012), p. 107.
10 Jean-Baptiste Jouffray et al., 'The Blue Acceleration: The Trajectory of Human Expansion into the Ocean', *One Earth*, 2 (2020), pp. 44–54.
11 Rachel Carson, *The Sea Around Us*, new edn (Oxford, 1961), p. v.
12 Alex Rogers, *The Deep: The Hidden Wonders of Our Oceans and How We Can Protect Them* (London, 2019), pp. 333–44; Roberts, *Ocean of Life*, pp. 318–31.

SELECT BIBLIOGRAPHY

Abulafia, David, *The Great Sea: A Human History of the Mediterranean* (London, 2012)
—, *The Boundless Sea: A Human History of the Oceans* (London, 2019)
Armitage, David et al., eds, *Oceanic Histories* (Cambridge, 2018)
Attlee, James, *North Sea: A Visual Anthology* (London, 2017)
Balsom, Erika, *An Oceanic Feeling: Cinema and the Sea* (New Plymouth, NZ, 2018)
Barkham, Patrick, *Coastlines: The Story of Our Shore* (London, 2015)
Bathurst, Bella, *The Wreckers: A Story of Killing Seas, False Lights and Plundered Ships* (London, 2005)
Bennett, Jonathan, *Around the Coast in Eighty Waves* (Dingwall, 2016)
Blass, Tom, *The Naked Shore: Of the North Sea* (London, 2015)
Burnett, John, *Dangerous Waters: Modern Piracy and Terror on the High Seas* (New York, 2003)
Carpenter, J. R., *An Ocean of Static* (London, 2018)
Carson, Rachel, *The Sea Around Us*, new edn (Oxford, 1961)
Clare, Horatio, *Down to the Sea in Ships: Of Ageless Oceans and Modern Men* (London, 2014)
—, *Icebreaker: A Voyage Far North* (London, 2017)
Cohen, Margaret, *The Novel and the Sea* (Princeton, NJ, 2010)
Compton, Nic, *Off the Deep End: A History of Madness at Sea* (London, 2017)
Corbin, Alain, *The Lure of the Sea: The Discovery of the Seaside in the Western World, 1750–1840*, trans. Jocelyn Phelps (Cambridge, 1994)
Cordingly, David, *Marine Painting in England, 1700–1900* (London, 1974)
Cornish, Vaughan, *Waves of the Sea and Other Water Waves* (London, 1910)
—, *Ocean Waves and Kindred Geophysical Phenomena* (Cambridge, 1934)
Cunliffe, Barry, *The Extraordinary Voyage of Pytheas the Greek* (London, 2001)

—, *On the Ocean: The Mediterranean and the Atlantic from Prehistory to AD 1500* (Oxford, 2017)

Deacon, Margaret, *Scientists and the Sea, 1650–1900: A Study of Marine Science*, 2nd edn (London, 1997)

Dear, I.C.B., and Peter Kemp, eds, *The Oxford Companion to Ships and the Sea*, 2nd edn (Oxford, 2005)

Doherty, Kieran, *Sea Venture: Shipwreck, Survival, and the Salvation of the First English Colony in the New World* (New York, 2007)

Dyer, Geoff, *Another Great Day at Sea: Life Aboard the USS George H. W. Bush* (Edinburgh, 2015)

Earle, Sylvia, *Sea Change: A Message of the Oceans* (New York, 1995)

—, *The World Is Blue: How Our Fate and the Ocean's Are One* (Washington, DC, 2009)

Ebbesmeyer, Curtis, and Eric Scigliano, *Flotsametrics and the Floating World* (New York, 2009)

Edwards, Philip, *The Story of the Voyage: Sea-narratives in Eighteenth-century England* (New York, 1994)

Fagan, Brian, *Beyond the Blue Horizon: How the Earliest Mariners Unlocked the Secrets of the Oceans* (London, 2012)

—, *The Attacking Ocean: The Past, Present and Future of Rising Sea Levels* (London, 2013)

Falconer, William, *Universal Dictionary of the Marine* (London, 1771)

George, Rose, *Deep Sea and Foreign Going: Inside Shipping, the Invisible Industry that Brings You 90% of Everything* (London, 2013)

Gilroy, Paul, *The Black Atlantic: Modernity and Double Consciousness* (London, 1993)

Gooley, Tristan, *How to Read Water: Clues, Signs and Patterns from Puddles to the Sea* (London, 2016)

Gray, Fred, *Designing the Seaside: Architecture, Society and Nature* (London, 2006)

Hamblyn, Richard, *Tsunami: Nature and Culture* (London, 2014)

Hamilton-Paterson, James, *Seven-tenths: The Sea and Its Thresholds*, 2nd edn (London, 2007)

Hattendorf, John B., ed., *The Oxford Encyclopedia of Maritime History*, 4 vols (Oxford, 2007)

Hay, David, *No Star at the Pole: A History of Navigation from the Stone Age to the 20th Century* (London, 1972)

Hoare, Philip, *Leviathan; or, the Whale* (London, 2008)

—, *The Sea Inside* (London, 2013)

—, *RisingTideFallingStar* (London, 2017)

Hobbs, Carl H., *The Beach Book: Science of the Shore* (New York, 2012)

Hohn, Donovan, *Moby-Duck: The True Story of 28,800 Bath Toys Lost at Sea* (London, 2012)

Klein, Bernhard, ed., *Fictions of the Sea: Critical Perspectives on the Ocean in British Literature and Culture* (Aldershot, 2002)

—, and Gesa Mackenthum, eds, *Sea Changes: Historicizing the Ocean* (London, 2004)

Langewiesche, William, *The Outlaw Sea: Chaos and Crime on the World's Oceans* (London, 2005)

Lavery, Brian, *The Conquest of the Ocean: The Illustrated History of Seafaring* (London, 2013)

Macfarlane, Robert, *The Wild Places* (London, 2007)

—, *Landmarks* (London, 2015)

Mack, John, *The Sea: A Cultural History* (London, 2011)

McKeever, William, *Emperors of the Deep: Sharks, The Ocean's Most Mysterious, Most Misunderstood, and Most Important Guardians* (London, 2019)

Mazzantini, Margaret, *Morning Sea: A Novel*, trans. Ann Gagliardi (London, 2015)

Myles, Douglas, *The Great Waves* (New York, 1985)

Nancollas, Tom, *Seashaken Houses: A Lighthouse History from Eddystone to Fastnet* (London, 2018)

O'Hanlon, Redmond, *Trawler: A Journey through the North Atlantic* (London, 2003)

Orsenna, Érik, *Portrait of the Gulf Stream: In Praise of Currents*, trans. Moishe Black (London, 2008)

Paine, Lincoln, *The Sea and Civilization: A Maritime History of the World* (London, 2014)

Parker, Bruce, *The Power of the Sea* (New York, 2010)

Pretor-Pinney, Gavin, *The Wavewatcher's Companion* (London, 2010)

Pye, Michael, *The Edge of the World: How the North Sea Made Us Who We Are* (London, 2014)

Raban, Jonathan, *Coasting: A Private Voyage* (London, 1986)

—, ed., *The Oxford Book of the Sea* (Oxford, 1992)

—, *Passage to Juneau: A Sea and Its Meanings* (London, 1999)

Rambelli, Fabio, *The Sea and the Sacred in Japan: Aspects of Maritime Religion* (London, 2018)

Riding, Christine, and Richard Johns, *Turner and the Sea* (London, 2013)

Roberts, Callum, *The Unnatural History of the Sea: The Past and Future of Humanity and Fishing* (London, 2007)

—, *Ocean of Life: How Our Seas Are Changing* (London, 2012)

Rogers, Alex, *The Deep: The Hidden Wonders of Our Oceans and How We Can Protect Them* (London, 2019)

Rothwell, Donald R. et al., eds, *The Oxford Handbook of the Law of the Sea* (Oxford, 2015)

Rozwadowski, Helen M., *Fathoming the Ocean: The Discovery and Exploration of the Deep Sea* (Cambridge, MA, 2005)

—, *Vast Expanses: A History of the Oceans* (London, 2018)

Runcie, Charlotte, *Salt on Your Tongue: Women and the Sea* (Edinburgh, 2019)

Ryan, William, and Walter Pitman, *Noah's Flood: The New Scientific Discoveries about the Event that Changed History* (New York, 1999)

Saylor, Eric, and Christopher M. Scheer, eds, *The Sea in the British Musical Imagination* (Woodbridge, 2015)

Schlee, Susan, *A History of Oceanography: The Edge of an Unfamiliar World* (London, 1975)

Strøksnes, Morten, *Shark Drunk: The Art of Catching a Large Shark from a Tiny Rubber Dinghy in a Big Ocean*, trans. Tiina Nunnally (London, 2017)

Thomas, Tony, *The Cinema of the Sea: A Critical Survey and Filmography, 1925–1986* (London, 1988)

Thompson, Christina, *Sea People: In Search of the Ancient Navigators of the Pacific* (London, 2019)

Thomson, David, *The People of the Sea: Celtic Tales of the Seal-folk* (Edinburgh, 2011)

Thomson, William, *The Book of Tides: A Journey Through the Coastal Waters of Our Island* (London, 2016)

Thoreau, Henry David, *Cape Cod*, ed. Joseph J. Moldenhauer (Princeton, NJ, 2004)

Urbina, Ian, *The Outlaw Ocean: Crime and Survival in the Last Untamed Frontier* (London, 2019)

White, Jonathan, *Tides: The Science and Spirit of the Ocean* (San Antonio, TX, 2017)

Whitworth, Victoria, *Swimming with Seals* (London, 2017)

Williams, Val, and Karen Shepherdson, *Seaside Photographed* (London, 2019)

Winton, John, ed., *The War at Sea, 1939–1945: An Anthology of Personal Experience* (London, 1967)

ASSOCIATIONS AND WEBSITES

International Council for the Exploration of the Sea
www.ices.dk/Pages/default.aspx

International Hydrographic Organization
https://iho.int/en/

Man in the Sea Museum
https://maninthesea.org

Marine Conservation Institute
https://marine-conservation.org

Maritime and Coastguard Agency
www.gov.uk/government/organisations/maritime-and-coastguard-
    agency

Mission Blue (Sylvia Earle Alliance)
https://mission-blue.org

National Oceanic and Atmospheric Administration (NOAA)
www.noaa.gov

National Oceanography Centre
https://noc.ac.uk

Oceanography Society
https://tos.org

Scripps Institution of Oceanography
https://scripps.ucsd.edu

Songs of the Sea
www.contemplator.com/sea

Woods Hole Oceanographic Institution
www.whoi.edu

ACKNOWLEDGEMENTS

My thanks are due to Daniel Allen and Michael Leaman for commissioning and editing this book, as well as to Amy Salter, Susannah Jayes and all of Reaktion's editors and designers for their careful work on the manuscript. It's a pleasure to thank the staff of the British Library, the Science Museum Library, Birkbeck Library and the Senate House Library, University of London, where much of this book was researched. I'm also grateful to the family, friends and colleagues with whom I have discussed and sometimes visited the sea over the years, including Jon Adams, Julia Bell, Lee Christien, Gregory Dart, John Drever, Markman Ellis, Angela Foster, Helen Frosi, David Hamblyn, Jonathan Kemp, Phil Mayo, Lena Müller, Michael Newton, Gavin Pretor-Pinney, George Revill, Ananda Rutherford, Ade Simpson and Colin Teevan.

My warmest thanks and love, though, are for Jo, Ben and Jessie Hamblyn, to whom this book is fondly dedicated.

PHOTO ACKNOWLEDGEMENTS

The author and publishers wish to express their thanks to the below sources of illustrative material and/or permission to reproduce it.

Alamy: pp. 11 (Chronicle), 91 (Pictorial Press Ltd), 99 (INTERFOTO), 157 (The Print Collector); Julia Bell: pp. 80, 192; Lee Christien: p. 36; CIA World Factbook: p. 114; Free Library of Philadelphia: p. 196; Getty Images: pp. 30–31 (Keystone-France); Richard Hamblyn: pp. 25, 28, 82 bottom, 84, 145, 147, 171, 186, 194, 204; From the collection of Richard Hamblyn: pp. 20, 41, 146, 153, 176, 183; *Harper's Weekly*; p. 95; Houghton Library, Harvard University: p. 12; Icarus Films: p. 148 (from *The Forgotten Space*, directed by Allan Sekula and Noël Burch, 2010); iStockphoto: p. 185 (benoitb); Library and Archives Canada: p. 101 (*Canadian Illustrated News*); Library of Congress: pp. 122, 125; Mary Evans Picture Library: p. 42 (Peter & Dawn Cope Collection); Metropolitan Museum of Art, New York: pp. 6 (Bequest of Julia B. Engel, 1984. Accession Number: 1984.341), 8 (Bequest of Miss Adelaide Milton de Groot (1876–1967), 1967. Accession Number: 67.187.207), 38 (The Elisha Whittelsey Collection, The Elisha Whittelsey Fund, 1959. Accession Number: 59.533.1539), 72 (Rogers Fund, 1919. Accession Number: JP1198), 92 (Catharine Lorillard Wolfe Collection, Wolfe Fund, 1906. Accession Number: 06.1234), 120 (Gift of Alexander Smith Cochran, 1913. Accession Number: 13.228.32), 150 (Purchase, John Osgood and Elizabeth Amis Cameron Blanchard Memorial Fund, Fosburgh Fund Inc. Gift, and Maria DeWitt Jesup Fund, 1971. Accession Number: 1971.192), 160 (Catharine Lorillard Wolfe Collection, Wolfe Fund, 1896. Accession Number: 96.29), 164 (Purchase, Dikran G. Kelekian Gift, 1922. Accession Number: 22.27.1), 165 (H. O. Havemeyer Collection, Gift of Horace Havemeyer, 1929. Accession Number: 29.160.35), 166 (Gift of George A. Hearn, 1906. Accession Number: 06.1281), 167 (Gift of George A. Hearn, 1910.

Accession Number: 10.64.5), 168 (H. O. Havemeyer Collection, Bequest of Mrs H. O. Havemeyer, 1929. Accession Number: JP1847), 169 (Fletcher Fund, 1926. Accession Number: 26.117), 170 (Gift of Alice E. Van Orden, in memory of her husband Dr. T. Durland Van Orden, 1988. Accession Number: 1988.353), 172 (Gift of John Goldsmith Phillips, 1976. Accession Number: 1976.646), 173 (The Elisha Whittelsey Collection, The Elisha Whittelsey Fund, 1972. Accession Number: 1972.567.13), 182 (Gift of Thomas Kensett, 1874. Accession Number: 74.29); NASA Earth Observatory: pp. 200–201; National Archives of the Netherlands (Nationaal Archief): pp. 21 (Spaarnestad Collection), 45 (Spaarnestad Collection); National Gallery, London: p. 162; National Maritime Museum, Greenwich, London: p. 143; National Museum of African American History & Culture (Smithsonian): p. 17; NOAA Photo Library: pp. 23 (Alaska ShoreZone Program NOAA/NMFS/AKFSC; Courtesy of Mandy Lindeberg, NOAA/NMFS/AKFSC), 55 (Archival Photography by Steve Nicklas, NOS, NGS), 58, 64 (Sean Lineham, NOAA, NGS, Remote Sensing), 67 (C&GS Season's Report Rigg 1926-69), 68, 75 (NOAA Central Library), 76 (NOAA Central Library), 94 (Okeanos Explorer), 96 (Julia Brownlee), 144 (NOAA's National Weather Service (NWS) Collection), 203 (NOAA Corps/Captain Budd Christman); Shutterstock: p. 49 (FoxPictures); Special Collections of the University of Amsterdam: pp. 56–7; Undersea Technologies/Undersea Labs (Habitats) & Obs: p. 177; University of Texas at Arlington Libraries: p. 53; University of Washington: p. 85 (Freshwater and Marine Image Bank); U.S. National Archives and Records Administration: p. 10 (National Archives at College Park/Still Picture Records Section, Special Media Archives Services Division (NWCS-S)); U.S. National Oceanic and Atmospheric Administration: p. 149; The Zuiderzee Museum, Netherlands: p. 107.

Chris Howells, the copyright holder of the image on pp. 14–15; Alessio Damato, the copyright holder of the image on p. 29; Cullen328, the copyright holder of the image on p. 113; and Elle plus at English Wikipedia, the copyright holder of the image on p. 118, have published them online under conditions imposed by a Creative Commons Attribution-Share Alike 3.0 Unported License. Flickr The Commons/ SPAARNESTAD PHOTO/Het Leven: p. 32; Michael Coghlan from Adelaide, Australia, the copyright holder of the image on p. 34, and Oregon State University, the copyright holder of the image on pp. 198–9, have published them online under conditions imposed by a Creative Commons Attribution-Share Alike 2.0 Generic License. This image was originally posted to Flickr by mikecogh at www.flickr.com/photos/89165847@N00/14283064095. It was reviewed on 17 March 2015 by FlickreviewR and was confirmed to be licensed under the terms of the CC-BY-SA-2.0. Brocken Inaglory, the copyright

INDEX

Page numbers in *italics* refer to illustrations